Support, movement and behaviour

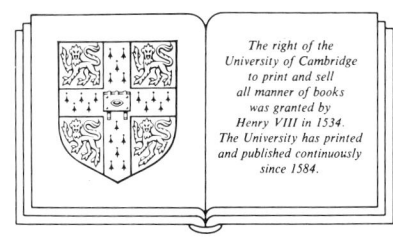

The right of the
University of Cambridge
to print and sell
all manner of books
was granted by
Henry VIII in 1534.
The University has printed
and published continuously
since 1584.

CAMBRIDGE UNIVERSITY PRESS

Cambridge
New York   *New Rochelle*
Melbourne   *Sydney*

Contents

Preface   v
How to use this unit   vi
Introduction to the unit   viii

Section 1 Movement and support   **1**
1.1 Introduction and objectives   1
1.2 Why is movement important?   1
1.3 The need for support   3
1.4 Plant movement   3
1.5 The principles of movement   4
1.6 Skeletons   4
1.6.1 The hydrostatic skeleton   6
1.7 Summary assignment 1   7
1.8 Past examination question   7

Section 2 Mechanical tissues and materials   **8**
2.1 Introduction and objectives   8
2.2 Skeletal materials – a review   8
2.3 Connective tissues and support   9
2.3.1 Areolar connective tissue   9
2.3.2 Dense fibrous connective tissue   10
2.3.3 Cartilage   10
2.3.4 Bone   11
2.3.5 Growth and development of bone   12
2.4 Summary assignment 2   13
2.5 Muscle   13
2.6 Mammalian support materials   15
Practical A: Examination of tissues involved in support   15
Extension practical B: Slide preparation of striated muscle   16
2.7 The ultrastructure of striated muscle   16
2.8 The chemistry and energetics of muscle contraction   18
2.8.1 Nervous stimulation of muscle fibre contraction   18

2.9 Turgor pressure and support in a young shoot   19
Practical C: Turgor pressure and support in plant tissue   20
2.10 Plant cells specialised for support   21
Extension practical D: Forensic investigations and plant tissues   22
2.10.1 The effect of the environment on plant support tissues   23
2.11 The insect cuticle   23
2.11.1 The insect cuticle and moulting   24
2.12 The advantages and disadvantages of an exoskeleton   25
2.13 Summary assignment 3   25
2.14 Past examination questions   25

Section 3 Forces, skeletons and joints   **26**
3.1 Introduction and objectives   26
3.2 Forces acting on the body   26
3.3 Stability and the centre of gravity   27
Practical E: Investigations on the length and position of legs in relation to stability   28
3.4 Supporting body weight   29
Practical F: Investigations on the strength of legs in relation to body weight   29
3.4.1 How bones resist environmental forces   30
3.4.2 Using braces to reduce stress   31
3.4.3 Counterbalancing   32
3.4.4 The role of cantilevers in mammalian support   32
3.4.5 The importance of arches in support   34
Practical G: Examining skeletons   34
3.5 Summary assignment 4   35
3.6 Vertebrate limbs   36
3.6.1 Artificial limbs   38
3.7 Joints   39
3.8 The importance of muscles and the nervous system in support   41
3.8.1 Coordination and control of skeletal muscle   42

3.9 Tendons 44
3.10 Pre-test: Levers 45
Practical H: Levers and their functions 45
3.10.1 Law of levers 46
3.10.2 Levers in the body 47
3.11 Plants and environmental forces 48
3.12 Summary assignment 5 48
3.13 Past examination questions 49

Section 4 Locomotion **51**
4.1 Introduction and objectives 51
4.2 Locomotion of Protista 51
Practical I: Investigating amoeboid movement 52
Practical J: Investigating ciliary and flagellar movement 53
4.3 Locomotion of the earthworm 54
Practical K: Investigating locomotion in an earthworm 54
4.4 Human locomotion – walking and running 55
4.4.1 Movement in some quadrupeds 57
4.4.2 Walking and the arthropods 59
4.5 Movement in water 60
4.5.1 The muscles of swimming 61
4.5.2 Stability in fish 61
4.5.3 Buoyancy in fish 61
4.5.4 Extension: Swimming in fish and other animals 62
4.6 Movement in air 63
4.6.1 Bird flight 63
Practical L: Flight in birds 65
4.6.2 Insect flight 66
4.7 Summary assignment 6 68
4.8 Extension: 'Muscle, a remarkable machine' 68
4.9 Past examination questions 68
4.10 Recommended reading for sections 1–4 69

Section 5 Behaviour **70**
5.1 Introduction and objectives 70
5.2 The importance of behaviour in mammals 71
AV 1: Behaviour for survival 71
5.3 The study of behaviour 72
5.3.1 Equipment for improving observation 75
5.3.2 The analysis of behaviour: Tinbergen and von Frisch 76
5.3.3 Laboratory behaviour: Pavlov, Skinner and Harlow 78

5.3.4 Observing and recording behaviour 81
Practical M: Exploration of a strange environment by rodents (open field box) 82
5.3.5 Extension: Physiological determinants of animal behaviour 83
5.3.6 Summary assignment 7 83
5.4 The classification of behaviour 83
5.4.1 Taxes and kineses 86
5.4.2 Investigation of some invertebrate responses in light 87
Practical N: Orientation in flatworms 87
Practical O: Response to light by fly larvae 88
Practical P: Response to light by woodlice and/or larvae of *Tenebrio molitor* (meal worm) 89
Practical Q: Orientation in brine shrimps (*Artemia* sp.) 90
5.4.3 Reflexes 90
5.4.4 Conditioned reflexes: classical conditioning 90
5.4.5 Operant or instrumental conditioning 90
5.4.6 Habituation 91
5.4.7 Trial-and-error learning 91
5.4.8 Latent learning 91
5.4.9 Imprinting 92
5.4.10 Insight learning or reasoning 93
5.4.11 Practical investigations of learning behaviour 94
Practical R: Withdrawal of tentacles by snails 94
Practical S: Typing in human beings 94
Practical T: Human learning with a pencil maze 95
5.4.12 The nature of memory 95
AV 2: The mechanisms of memory 95
5.4.13 Summary assignment 8 96
5.4.14 Extension: The evolution of intelligence 96
5.5 The internal environment 97
5.5.1 Hormones and behaviour 97
5.5.2 Biological rhythms (endogenous rhythms) 98
5.5.3 Photoperiodism 99
5.6 Communication 100
5.6.1 Communication in honey-bees 100
5.6.2 Pheromones 101
5.6.3 Song 102
5.6.4 Courtship 102
5.6.5 Extension: The curious behaviour of the stickleback re-examined 103
5.7 Social behaviour 104
5.8 The evolution of behaviour 104

5.8.1 Cross-species studies 104
5.8.2 Genes and behaviour 106
5.9 A case-study in behaviour: The male Siamese fighting fish 107
AV 3: The display of the Siamese fighting fish 107
5.10 Summary assignment 9 107
5.11 Past examination questions 108
5.12 Recommended reading 109

Section 6 Self tests 110
Self test 1 110
Self test 2 110
Self test 3 111
Self test 4 112
Self test 5 114
Self test 6 114
Self test 7 115

Section 7 Answers to self tests 116
Self test 1 116
Self test 2 116
Self test 3 117
Self test 4 118
Self test 5 120
Self test 6 121
Self test 7 123

Section 8 Answers to self-assessment questions 124
Pre-test: Levers 132

Index 133

Published by the Press Syndicate of the University of Cambridge
The Pitt Building, Trumpington Street, Cambridge CB2 1RP
32 East 57th Street, New York, NY 10022, USA
10 Stamford Road, Oakleigh, Melbourne 3166, Australia

© Cambridge University Press 1987

First published 1987

Printed in Great Britain at the University Press, Cambridge

British Library cataloguing in publication data
Support, movement and behaviour.—(ABAL; unit 7)
  1. Biomechanics   2. Locomotion
  I. Series
  574.1′8   QP301
  ISBN 0 521 28830 4

Preface

The Inner London Education Authority's Advanced Biology Alternative Learning (ABAL) project has been developed as a response to changes which have taken place in the organisation of secondary education and the curriculum. The project is the work of a group of biology teachers seconded from ILEA secondary schools. ABAL began in 1978 and since then has undergone extensive trials in schools and colleges of further education. The materials have been produced to help teachers meet the needs of new teaching situations and provide an effective method of learning for students.

Teachers new to A-level teaching or experienced teachers involved in reorganisation of schools due to the changes in population face many problems. These include the sharing of staff and pupils between existing schools and the variety of backgrounds and abilities of pupils starting A-level courses whether at schools, sixth form centres or colleges. Many of the students will be studying a wide range of courses, which in some cases will be a mixture of science, arts and humanities.

The ABAL individualised learning materials offer a guided approach to A-level biology and can be used to form a coherent base in many teaching situations. The materials are organised so that teachers can prepare study programmes suited to their own students. The separation of core and extension work enables the academic needs of all students to be satisfied. Teachers are essential to the success of this course, not only in using their traditional skills but for organising resources and solving individual problems. They act as personal tutors, and monitor the progress of each student as he or she proceeds through the course.

The materials aim to help the students develop and improve their personal study skills, enabling them to work more effectively and become more actively involved and responsible for their own learning and assessment. This approach allows the students to develop a sound understanding of fundamental biological concepts.

Acknowledgements

Figures: 1*a*, 1*e*, 2*a*, 4, 6, 10, 11, 40, 43, 44*a*, 44*b*, 44*d*, 49, 68, 74*a*, 74*b*, 110, A. Langham; 1*b*, 1*c*, Popperfoto Ltd; 1*d*, V. Kiernan; 2*b*, 8, 9, 15, 16, 17, 18, 19*a*, 19*b*, 23, 24, 25, 27, 36, 38, 96*c*, 119*a*, Biophoto Associates; 2*c*, Royal Botanic Gardens, Kew, copyright 1986; 2*d*, 54, 126, 158 (fox), 166*a*, 166*b*, Natural History Photographic Agency; 3, from *Scientific American* offprint 1382, vol. 238, no. 2, based on an original drawing by Yolande Heslop Harrison; 5, Horniman Museum, London; 7, from *Sports Science* by R. Hawkey (1981) Hodder & Stoughton Educational; 30*a*, by kind permission of Dr A. Elliott; 30*b*, 30*c*, based on a graphic by Colin R. Hopkins; 32, U.J. McMahan, N.C. Spitzer & K. Peper (1972) *Proc. Roy. Soc. Lond. (B)*, **181**, 421–30, by kind permission of the Royal Society; 39, by kind permission of Marcus Barber; 44*c*, J. Allan Cash Ltd; 50, 52, R. McNeill Alexander *Animal Mechanics*, Sidgwick & Jackson; 67, J.E. Hanger & Co. Ltd., Roehampton; 78, G. Hardin & C. Bajema *Biology: Its Principles and Implications*, 3rd ed., W.H. Freeman & Company; 92, by kind permission of Manchester Museum; 101, Gray & Lissman (1938) *J. Exp. Biol.* **15**, 506–17; 105, D. McKean *Introduction to Biology*, fig. 24–11, p.124; 112, James Gray (1953) *How Animals Move*, Cambridge University Press; 115, Dr Peter Whitehead (1975) *How Fishes Live*, Elsevier Copyrights Management S.A.; 120, from *Biomechanics Minicourse Development Project* (1976) Saunders College Publishing, USA; 127, by kind permission of H. Kacher; 128, by kind permission of R.T. Hutchings; 131, 132, N. Tinbergen, *Social Behaviour in Animals*, Chapman & Hall; 133, reproduced with permission from the *Annual Review of Entomology*, vol.1 © 1956 by Annual Review Inc., *The Language and Orientation of the Honey Bee*, K. von Frisch & M Lindauer; 134, by kind permission of Don Briggs, Digby Stuart College; 135, University of Wisconsin Primate Laboratory, USA; 136, D. Lack, *The Life of the Robin*, H.F. & G. Witherby Ltd.; 139, V.G. Dethier & Eliot Stellar (1970) *Animal Behaviour*, p.91, reprinted by permission from Prentice-Hall Inc., Englewood Cliffs, New Jersey, USA; 141, Ullyot, *J. exp. Biol.*, 1936; 142, D.L. Gunn *J. exp. Biol.*, 1937; 151, P.J.B. Slater (1978) *Sex Hormones and Behaviour*, Studies in Behaviour series, Edward Arnold; 152, M.O.K. Jones, M. Hall & A.M. Hope (1967) *J. exp. Biol.* **47**, by kind permission of the Company of Biologists Ltd.; 153, B. Lofts, B.K. Follett & R.K. Murton (1970) *Soc. Endocr.* **18**, p.545; 158 (gull), Heather Angel Photo Library; 160, James Gould *Ethology: The Mechanisms and Evolution of Behaviour*; 162, Ardea.

Text: pp.73–4, K.Z. Lorenz (1937) Imprinting, *Auk*, **54**, 245–73; pp.78–9, I.P. Pavlov *Conditioned Reflexes*, Dover Publications Inc. N.Y.; pp.80–1, Harry Harlow, Love in infant monkeys, *Scientific American* June 1959.

Examination questions: By permission of the University of Cambridge Local Examinations Syndicate, the University of London, University Entrance and School Examinations Council, the Associated Examining Board, the Oxford & Cambridge Schools Examination Board and the Joint Matriculation Board.

How to use this unit

This is not a textbook. It is a guide that will help you learn as effectively as possible. As you work through it, you will be directed to practical work, audio-visual resources and other materials. There are sections of text in this guide which are to be read as any other book, but much of the guide is concerned with helping you through activities designed to produce effective learning. The following list gives details of the ways in which the unit is organised.

(1) Objectives

Objectives are stated at the beginning of each section. They are important because they tell you what you should be able to do when you have finished working through the section. They should give you extra help in organising your learning. In particular, you should check after working through each section that you can achieve all the stated objectives and that you have notes which cover them all.

(2) Self-assessment questions (*SAQ*)

These are designed to help you think about what you are reading. You should always write down answers to self-assessment questions and then check them immediately with those answers given at the back of this unit. If you do not understand a question and answer, make a note of it and discuss it with your tutor at the earliest opportunity.

(3) Summary assignments

These are designed to help you make notes on the content of a particular section. They will provide a useful collection of revision material. They should therefore be carried out carefully and should be checked by your tutor for accuracy. If you prefer to make notes in your own way, discuss with your tutor

whether you should carry out the summary assignments.

(4) Self tests

There are one or more self tests for each section. They should be attempted a few days after you have completed the relevant work and not immediately after. They will help you identify what you have not understood or remembered from a particular section. You can then remedy any weaknesses identified. If you cannot answer any questions and do not understand the answers given, then check with your tutor.

(5) Tutor assessed work

At intervals through the unit you will meet an instruction to show work to your tutor. This will enable your tutor to monitor your progress through the unit and to see how well you are coping with the material. Your tutor will then know how best to meet your individual needs.

(6) Past examination questions

At various points in the unit you will come across past examination questions. These are only included where they are relevant to the topic under study and have been selected both to improve your knowledge of that topic and also to give you practice in answering examination questions.

(7) Audio-visual material

A number of activities in this unit refer to video sequences which may be available from your tutor. They deal with topics which cannot be covered easily in text or practical work as well as providing a change from the normal type of learning activities. This should help motivate you.

(8) Extension work

This work is provided for several reasons: to provide additional material of general interest, to provide more detailed treatment of some topics, to provide more searching questions that will make demands on your powers of thinking and reasoning.

(9) Practicals

These are an integral part of the course and have been designed to lead you to a deeper understanding of the factual material. You will need to organise your time with care so that you can carry out the work suggested in a logical sequence. If your A-level examination requires your practical notebook to be assessed, you must be careful to keep a record of this work in a separate book. A hazard symbol, ☠ , is used in the Materials and Procedures sections to mark those substances and procedures which must be treated with particular care.

(10) Discussions

Talking to one another about biological ideas is a helpful activity. To express yourself in your own words, so that others can understand you, forces you to clarify your thoughts. When a sufficient number of your class (at least three, but not more than five) have covered the material indicated by a discussion instruction, you should have a group discussion. Question individuals if what they say is not clear. This is the way you will both learn and understand.

(11) Post-test

A post-test is available from your tutor when you finish this unit. This will be based on past examination questions and will give you an idea of how well you have coped with the material in this unit. It will also indicate which areas you should consolidate before going on to the next unit.

Study and practical skills

The ABAL introductory unit *Inquiry and investigation in biology* introduced certain study and practical skills which will be practised and improved in this unit. These included
(a) the QS3R method of note-taking;
(b) the construction of graphs, histograms and tables;
(c) the analysis of data;
(d) drawing of biological specimens;
(e) use of the light microscope;
(f) the design of practical investigations;
(g) comprehension of written reports;
(h) discussion groups.

Pre-knowledge for this unit

This unit has been written to follow on from the unit *Response to the environment* and assumes the knowledge contained within it. Thus, you need a knowledge of receptor organs, the mode of functioning of nerves and hormones and the coordination of responses by the central nervous system. An understanding of the variety of plant responses and of reproduction in animals is also required.

Introduction to the unit

This unit is essentially a sequel to unit 6 *Response to the environment*, and focuses on the final stage of the process that begins with the reception by an organism of a stimulus from its environment, that is, the response by an effector. This effector may be a muscle or gland or, in the case of plants, a directed change in its biochemistry. One of the commonest categories of response is for movement to occur and for this some sort of support system is required.

The first three sections of this unit are concerned with an understanding of how support and movement are brought about in living organisms. This involves a study of skeletal materials, muscle tissue and systems and the environmental forces which act upon plants and animals. Section 4 is concerned with some of the ways in which animals are adapted to move around in water, on land and through the air.

The various movements of an animal, movements of parts or of the whole of its body, constitute what we can actually observe of its reactions to the environment. These movements, adapted towards the animal's survival and usually reversible and repeatable, are what we describe as behaviour. This is a rapidly growing area of biological study and the final section of this unit can only attempt to introduce this topic and give something of the 'flavour' of the various approaches to the study of what animals actually do in their daily lives.

Section 1 Movement and support

1.1 Introduction and objectives

A child walking along a beach and seeing a jellyfish or crab cast up by the tide will often poke it to decide if it is alive or dead. If some movement results it is thought to be still living, for this is perhaps the most obvious characteristic of living organisms. Movement is a common response to a variety of stimuli and an important survival mechanism which can adapt the organism to environmental changes.

If the whole or part of an organism is to be moved, some kind of support system is usually involved. Support provides organisms with a definable shape, enables them to react against forces and to exert a force against the medium in which they must move. Most animals and protists are capable of **locomotion**, moving the whole of their bodies from place to place, but some are **sessile**, being attached at their base to a substrate and capable only of moving parts of their body.

Plant movement is usually restricted to organs such as stem tips, leaves and flowers. Male reproductive cells of the lower plants are often capable of locomotion, and seeds show many adaptations for being passively moved by wind, water and other organisms. Stomata open and close and many flowers show daily movements.

This section covers the reasons for movement, the connections between support and movement, the need for a framework or skeleton and studies the variety of plant movements.

After completing this section you should be able to do the following.

(*a*) Distinguish between movement and locomotion.

(*b*) State the reasons why movement is important to living organisms.

(*c*) Give the reasons why living organisms require some means of support.

(*d*) Give an account of movement in plants.

(*e*) List the various types of 'skeleton' to be found in living organisms.

(*f*) State the fundamental difference in support mechanisms of plants and animals.

1.2 Why is movement important?

The photograph in figure 1 shows five different actions performed by people. Study these carefully and then answer the question below.

1   Actions involving movement

(a)

(c)

(b)

(d)

(e)

SAQ 1 State the action illustrated by each photograph and suggest a possible function for each.

You can probably think of many other types of movement carried out by human beings and other animals. Some examples include breathing movements to ventilate the lungs, the rhythmic movements of the heart which maintain the flow of blood through the body, the movements of animals in search of a mate prior to reproduction and the long-distance movements or migrations of certain animals to avoid unfavourable seasons.

Movement in plants is often very slow and is best observed by the use of time-lapse photography. Figure 2 shows four examples of movement in plants.

Study the photographs carefully and then answer the question below.

SAQ 2 For the movements illustrated in each photograph, suggest an advantage to the plants concerned.

SAQ 3 Suggest a major reason why most animals have a need for frequent locomotion and plants do not.

SAQ 4 Name three habitats in which animals can live a sessile existence and explain how this mode of life is possible.

2 Movement in plants

(b)

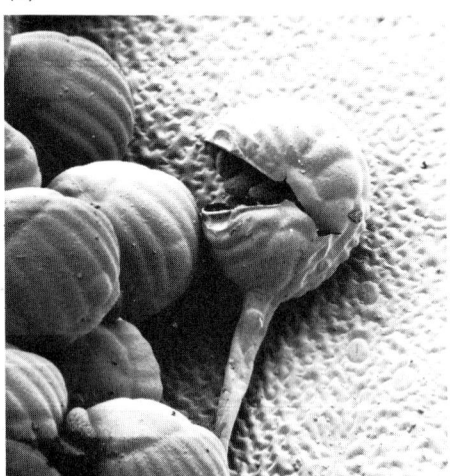

(a)

(c)

(d) Nocturnal opening of flower

1.3 The need for support

Whether you just observe or touch and feel animals and plants it will be evident that they have a shape and form. This shape and form is created by the effect of outer binding tissues which hold the internal contents of the organisms in place. Some aquatic organisms are so fragile that, although they have an outer supporting layer, their shape and form is only fully maintained by the extra support gained from the aquatic environment in which they live, for example some jellyfish and hydroid colonies, and plants such as the water buttercup and seaweeds. When the support from the water is removed the characteristic shape of the organism is no longer clearly displayed.

Green plants produce the main materials they need for metabolism and growth by photosynthesis and they must be supported in such a way that their leaves are held out to receive adequate light energy. Similarly, sedentary filter feeders need support for their feeding apparatus if it is to trap small particles.

For an organism to actively move itself from one place to another, it needs some means of exerting a push or pull against the medium in or on which it must move. Organs of locomotion need a means of firm support.

1.4 Plant movement

If green plants are exposed to a unilateral light source they show a movement towards this light source. This response, along with many others, illustrates that plants, which in general are sedentary photosynthesisers, respond to a variety of stimuli by undergoing relatively rapid growth activities. These growth movements are called **tropisms** and the growth pattern is controlled by plant hormones. A detailed account of tropic movements in response to light and gravity is given in the unit *Response to the environment*.

SAQ 5 Define a tropic movement.

Two other examples of tropism are **hydrotropism**, by which roots grow towards moisture (though there is some doubt as to the existence of this phenomenon), and **chemotropism**, in which a chemical stimulates growth. An example of chemotropism is the growth of a pollen tube towards the ovary in a flowering plant. The reaction of a plant part such as a tendril to a touch stimulus is called **thigmotropism** if the direction of the response is aligned with the stimulus, or **thigmonasty** if it is not directional. (Nastic movements are also discussed in *Response to the environment*.) Rapid responses in tendrils are thought to be brought about by movements of ions. Recent experiments show that rapid changes in ATP and inorganic phosphate content take place after pea tendrils are touched. It may be that the permeability of the membranes alters, or else that active transport of ions occurs.

SAQ 6 How would the latter explanation account for changes in ATP and inorganic phosphate levels?

SAQ 7 What name is given to a nastic movement occurring in response to a change in light intensity?

Some plants such as tulips and crocus show repeated opening and closing movements of the flowers in response to temperature changes as small as a fraction of a degree. These **thermonastic** movements are permanent growth movements resulting from a growth differential between the upper and lower tissues of petals. The mechanism is not known.

A number of insectivorous plants are equipped with traps that react rapidly enough to catch live insects. Some of these combine rapid movements with a special trigger device. The Venus fly-trap video sequence is studied in units *Inquiry and investigation* and *Response to the environment*. The bladderwort, *Utricularia* sp., an aquatic plant, has spherical bladders which are under pressure and opened by a trapdoor arrangement. When a small arthropod swims to the bladder and touches a trigger hair the door opens rapidly inward. The organism is swept inside by the movement of the door and by water entering the bladder by suction. The trap is then closed and reset. Figure 3 illustrates this trapping action.

3 Trapping action of *Utricularia*

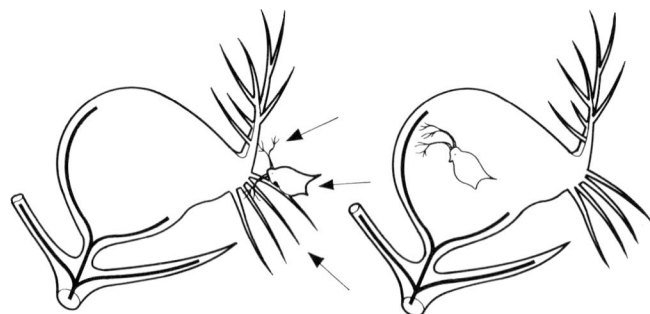

The insect touches a trigger hair and trips the trap, which opens and engulfs the insect. The trap is then reset.

The leaves of many plants undergo 'sleep movements' a rhythmic raising of leaves in the morning and folding of them in the evening.

SAQ 8 What technical name is given to such movements? Name a plant which shows such movements of its leaves.

1.5 The principles of movement

Within our solar system, the planets are continuously orbiting around the Sun. They are able to do this because they are moving through an almost complete vacuum. The planets are thus able to keep to a constant speed. For things moving on the Earth, two factors can affect movement by causing a slowing down or drag effect. These are friction against the surface being moved upon, and resistance (from air and water) of the medium being moved through. Gravity also affects movement by acting as a downward force on the body of the mover. Movement requires that these factors be overcome and, to do this, an organism requires energy.

SAQ 9 (*a*) What is the source of this energy for an animal?
(*b*) During which metabolic process is it made available?

In order to move the animal, the potential energy of the system has to be transformed into kinetic energy. The majority of animals have moving parts, working in conjunction with muscles, which enable them to change their position by exerting a thrust against the surrounding medium. Arthropods and vertebrates possess jointed appendages or limbs, earthworms use their chaetae, and swimming organisms from many groups use the movement of their flexible body, tail or flagellum to propel them through water. To achieve locomotion an animal has to expend energy to exert a downward and backward force on the substrate to act against the opposing forces of resistance, friction and gravity.

Finally, since animals, unlike planets, do not usually move continuously, there must be something which initiates the movement and something which stops it. There must be a control mechanism which coordinates all the activities of an animal which allows it to move.

SAQ 10 Which system of the body is likely to be involved in control?

SAQ 11 List the basic principles of movement in animals.

1.6 Skeletons

The cell is the building block of living organisms. There are significant differences between the cells of plants and animals. This has led to differences in the design of their support systems.

Most plant cells are surrounded by a rigid secondary cell wall outside the cell membrane. This supporting structure has been adapted for the support system of the whole organism. Animal cells are surrounded by a cell membrane alone. This membrane is not a rigid structure. Hence, the support system in animals is based on **extracellular materials** secreted by cells rather than structures of the cell itself.

Long after the death of an organism, certain parts of its body often remain intact, as the photographs in figures 4, 5 and 6 show. These are parts of the skeletal system which, in life, give support to the organism.

4 Coral (skeletal elements)

5 Leaf skeleton

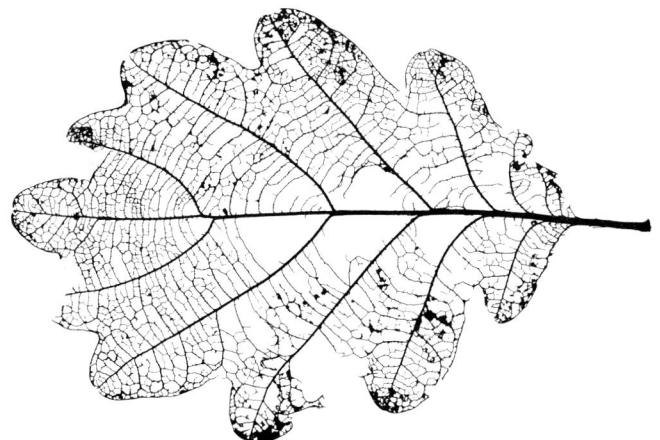

6 Prehistoric skeleton in grave

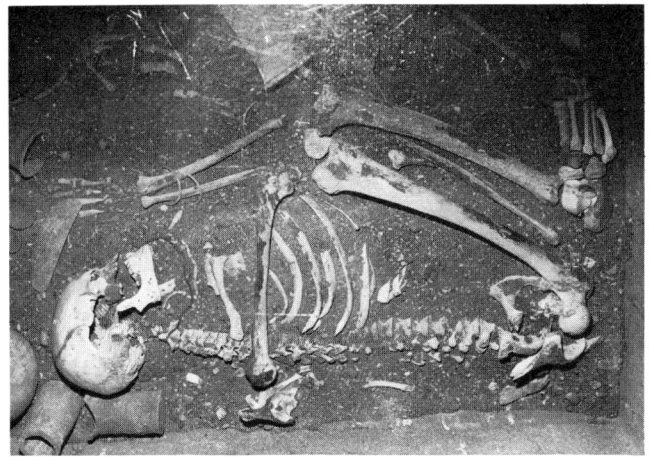

Figure 7 shows an artist's impression of the human body without its support system. Notice how the absence of the support system reduces the body to a shapeless mass. The skeletal system is important in maintaining shape and supporting the body.

7 Artist's impression of a human body without a skeleton

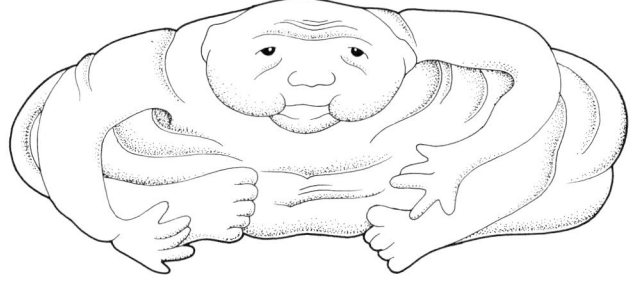

SAQ 12 What are the other functions of a skeleton?

Skeletons may be external (**exoskeletons**) or internal (**endoskeletons**). Shells are one common form of exoskeleton found in several invertebrate groups and in the Protista. Radiolarians build an internal skeleton of silica, while the external calcareous shells of dead foraminiferan protozoans are responsible for large chalk deposits such as the cliffs of Dover. These are shown in figures 8 and 9. The shelled amoeba *Difflugia* cements sand grains together, but the ciliophora are covered by a living pellicle formed of a number of membranes.

8   Radiolarians

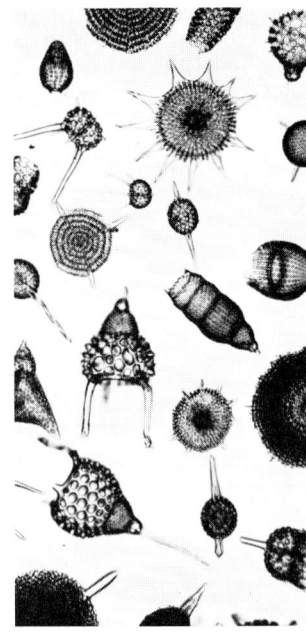

9   Foraminiferans

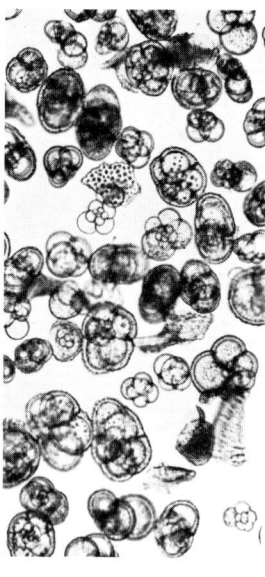

11   Mollusc shells

Each individual coral polyp secretes its own chalky skeleton. In life this is clothed with a translucent film of living tissue and these colonial animals are responsible for building reefs and islands in shallow, tropical seas. An example of a non-living coral skeleton was shown in figure 4. Figure 10 shows a sea urchin test made from fused plates of calcium salts called ossicles which lie just under the skin of a living urchin and bear long spines. Figure 11 shows a collection of mollusc shells, including the internal shell of a cuttlefish which also serves the animal as a buoyancy device and allows the animal to swim at different depths.

10   Sea urchin test

The cuticle which forms the exoskeleton of arthropods sends processes into the living tissues for muscle attachment. Skeletons which function in movement as well as providing protection, work in conjunction with muscle tissue.

The contraction of muscle fibres is an active process but their relaxation is not. An external force must be applied to restore the fibres to their original length. In some animal groups, this force is applied through a jointed skeleton to which muscles are attached in such a way that the contraction of one muscle brings about the relaxation of another. Such muscles are referred to as **antagonistic pairs**.

1.6.1 The hydrostatic skeleton

Jointed skeletons appeared relatively late in evolutionary history. From early times, animals exploited a **hydrostatic skeleton**. Water has three very important properties:
(*a*) it is incompressible;
(*b*) it will transmit pressure changes equally in all directions;
(*c*) it has a low viscosity and is easily deformed.

Thus, water or body fluids when contained in a closed system may provide a 'skeleton' which animals (and plants) can use for support and movement.

12 Animal with circular muscles only

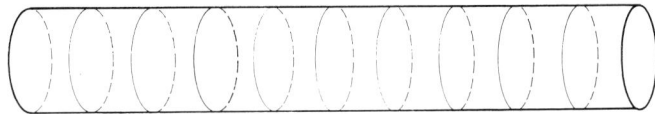

SAQ 13 In an animal with a hydrostatic skeleton (such as that shown in figure 12), what will happen if the circular muscles at the left-hand side contract when (*a*) the body can extend, (*b*) the body is not free to extend?

SAQ 14 In which situation can contraction of the left-hand muscles bring about relaxation of the muscles on the right-hand side?

SAQ 15 Suggest an additional set of muscles which could bring about relaxation of the circular muscles whether or not the body is free to extend.

This basic model can be used to explain movement in a simple animal such as a sea anemone. Study figure 13.

13 Plan of a sea anemone

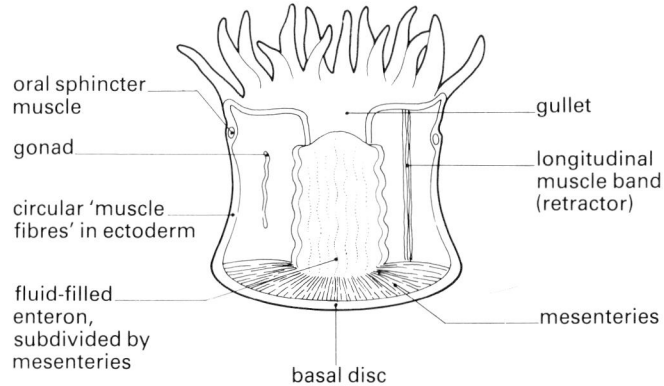

SAQ 16 What occupies the space of the enteron?

SAQ 17 What is the effect of closing the sphincter muscle?

SAQ 18 Coelenterates do not possess true muscles.
(*a*) What is the name given to the contractile elements in this group?
(*b*) Name the antagonistic 'muscles'.

SAQ 19 Explain concisely how a sea anemone might bend to the right, using your answers to SAQs 16–18.

Hydrostatic skeletons are found in a variety of soft-bodied invertebrates. Locomotion in the earthworm, which is dependent on the presence of a hydrostatic skeleton, is studied in section 4 of this unit.

Water in the vacuoles of plant cells forms a type of hydrostatic skeleton and provides much of the support in herbaceous plants, leaves and flowers. Changes in turgor pressure make possible movements such as the opening and closing of stomata, the response of the leaflets of the sensitive plant, opening and closing of flowers in response to changing levels of illumination and various 'sleep' movements.

Herbaceous plants are supported by a combination of turgor pressure and rigid secondary cellulose cell walls. **Fibres** are elongated cells found in the phloem and xylem. Their secondary walls are very thick, with cellulose microfibrils orientated in a lengthwise direction. The function of these cells is mechanical support. The xylem also functions as a support tissue. A tree is basically held up by its wood – the lignified cells of the xylem tissue. The role of turgor pressure in plant support is investigated in section 2, practical C.

1.7 Summary assignment 1

Re-read the objectives in section 1.1 of this unit. Under the headings (*a*) to (*f*) make brief notes fulfilling the tasks described in the objectives.

Self test 1, page 110, covers section 1 of this unit.

1.8 Past examination question

Discuss fully the importance of turgor changes and tropic responses to plants. (Detailed descriptions of the mechanisms of turgor changes and tropic responses are not required.)

(University of Cambridge Local Examinations Syndicate, 1979)

Section 2   Mechanical tissues and materials

2.1 Introduction and objectives

This section will familiarise you with the variety of organic and inorganic materials which provide support for living organisms. Certain tissues are specially adapted to bear mechanical stress and to act as support within organisms; their structural properties and mode of functioning is studied.

After completing this section you should be able to do the following.

(*a*) Detail the role played by collagen in animal support tissues.

(*b*) Give examples of other organic compounds used for support in living organisms.

(*c*) Recognise and describe the following animal support tissues: connective tissue, cartilage, bone and muscle.

(*d*) Distinguish between cartilage and membrane bones and outline their formation.

(*e*) Describe the ultrastructure of striated muscle and relate it to its method of contraction.

(*f*) Describe the structure and functioning of neuromuscular junctions.

(*g*) Give an account of the role of turgor pressure, collenchyma, sclerenchyma and xylem fibres in the support of plants.

(*h*) Describe how the structure of an insect's cuticle adapts it to function efficiently.

(*i*) State the advantages and disadvantages of an exoskeleton.

2.2 Skeletal materials – a review

Most animal skeletons are made of proteins such as **collagen**, which is hardened by the deposition of mineral salts or toughened by chemical processes similar to the tanning of leather or the hardening of rubber. Figure 14 indicates the ways in which collagen may be transformed into skeletal material. Study this diagram.

14   Skeletal materials based on collagen

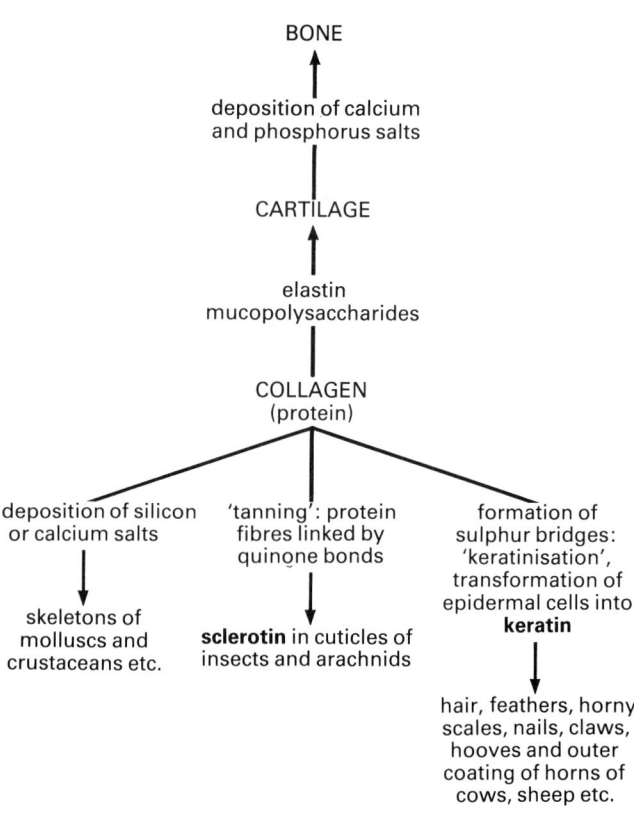

SAQ 20 Name two vertebrate skeletal materials based on collagen.

SAQ 21 What is the end-product of the tanning of collagen? Which animal group possesses a tanned protein in their exoskeleton?

In addition to being incorporated into protein structures such as cartilage, polysaccharides with long fibrous molecules of considerable mechanical strength form skeletal materials. **Chitin** is a nitrogen-containing polysaccharide present in the cuticle of insects.

SAQ 22 Name a polysaccharide which is a support material for plants.

The 'wood' of trees and bushes is formed by the xylem tissue and provides mechanical support. Xylem cells have thickened walls formed by the deposition of **lignin** which makes them strong and rigid. Lignin is a complex **aromatic** substance (containing a benzene ring) the structure of which is still not completely known.

2.3 Connective tissues and support

Connective tissues support and hold together the various tissues and organs of an animal's body. These tissues serve as a connecting system, forming sheaths around organs, bundles in which lie nerves and blood vessels, sheets which attach the skin to underlying tissues and the linings of the various body cavities. In vertebrates, the name 'connective' tissue is extended to include the skeletal tissues and blood since a common basic structure is found in them all.

(a) They have a matrix of non-living material in which cells are contained.
(b) These cells have secreted the matrix substance.
(c) They possess *fibres* running in the matrix.

Developmental studies of vertebrates show that all connective tissues have a common origin from one type of embryonic cell.

SAQ 23 Apart from its embryonic origin, blood does not seem to be a typical connective tissue. Which characteristic of connective tissue is shown by blood?

Practical A will help you to relate your knowledge of the various connective tissues to actual specimens and prepared slides.

2.3.1 Areolar connective tissue

Areolar connective tissue is the most widely distributed connective tissue, surrounding and penetrating organs such as muscles, nerves, bones and tendons. It provides the external covering of kidneys and liver and forms the mesenteries of the abdomen and the meninges of the brain.

15   Areolar connective tissue × 420

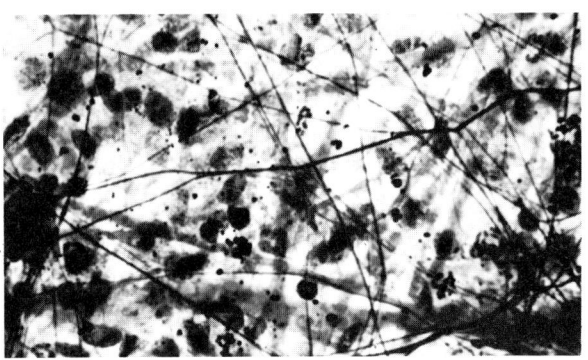

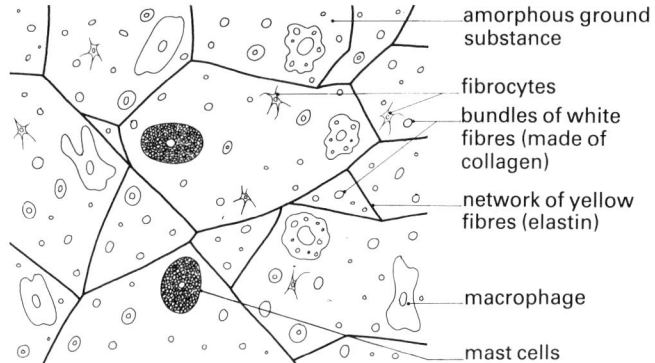

amorphous ground substance
fibrocytes
bundles of white fibres (made of collagen)
network of yellow fibres (elastin)
macrophage
mast cells

Study the photograph in figure 15 and its accompanying diagram.

SAQ 24 The ground substance of areolar tissue is gelatinous and has cells and fibres embedded in it. List (a) the types of fibre, and (b) the varieties of cells.

Collagen fibres occur in bundles and form wavy strands. They have great tensile strength and yet they are flexible. This allows tissues containing them to resist deformation by stretching.

Elastic fibres are fine, branching elastin fibres with one-fiftieth of the tensile strength of collagen. However, they can extend to double or more their length and then return to their resting length. They are important in tissues which need to stretch.

Both these types of fibre provide support in the walls of blood vessels.

SAQ 25 Suggest a reason why more elastic fibres are found in artery walls than vein walls.

The fibres are produced by cells known as **fibroblasts**. These cells are the living and dividing cells (basically they are immature cells) which also produce the matrix of the tissue. The **matrix** (ground tissue) is largely composed of a solution of mucopolysaccharides. When cells storing fat predominate in the tissue it is called **adipose** tissue. This functions as packing tissue and as a metabolic reserve.

2.3.2 Dense fibrous connective tissue

Tendons are formed by dense fibrous connective tissue. A tendon attaches muscle to bone. Study figure 16.

SAQ 26 List the ways in which this connective tissue differs from areolar connective tissue.

SAQ 27 Suggest a reason why elastic fibres are absent from tendons.

Ligaments which join bone to bone, are similar in structure to tendons but the fibres are less regularly arranged.

2.3.3 Cartilage

Cartilage cells, or **chondrocytes**, have a rounded shape and lie in little spaces within a matrix of mucopolysaccharides (figure 17). Within this sponge-like mesh of fibrils are the negatively charged ions of chondroactin sulphate and keratin sulphate. The negatively charged sulphate ions attract positive ions, especially sodium. These ions lower the water potential, drawing water in, and this gives the cartilage its stiffness. Cartilage is better at resisting squeezing and stretching than other connective tissues, but it is not as strong as bone.

17 Hyaline cartilage

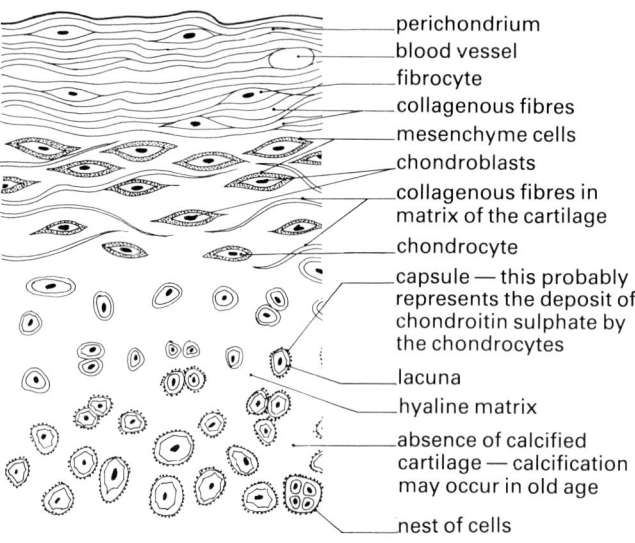

perichondrium
blood vessel
fibrocyte
collagenous fibres
mesenchyme cells
chondroblasts
collagenous fibres in matrix of the cartilage
chondrocyte
capsule — this probably represents the deposit of chondroitin sulphate by the chondrocytes
lacuna
hyaline matrix
absence of calcified cartilage — calcification may occur in old age
nest of cells

16 LS human tendon × 320

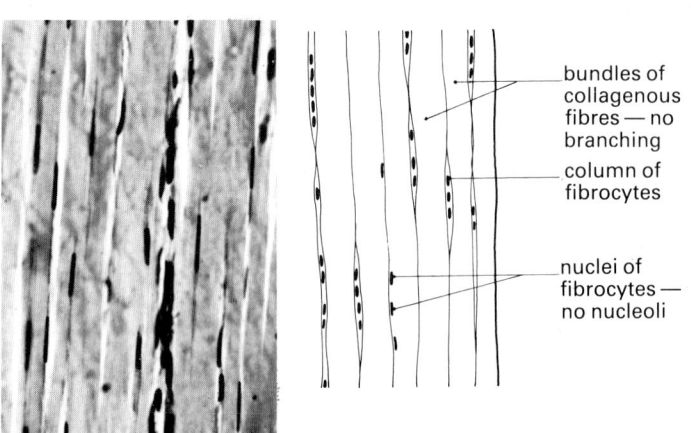

bundles of collagenous fibres — no branching

column of fibrocytes

nuclei of fibrocytes — no nucleoli

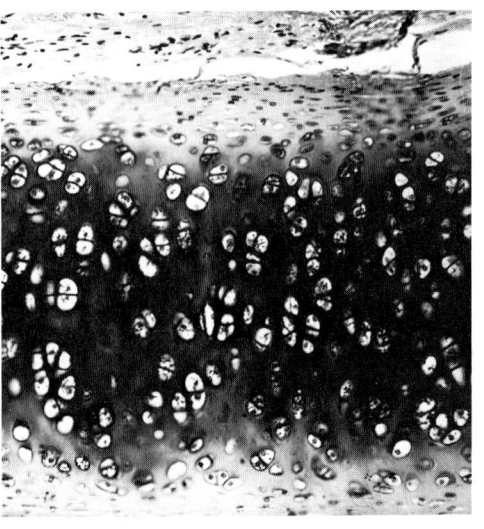

Collagen and elastin fibres may be present in varying amounts.

SAQ 28 Look at figure 17. What component of connective tissue is missing from hyaline cartilage?

Cartilage forms the skeletal material of the Chondrichthyes.

SAQ 29 Name two animals belonging to this class.

Cartilage provides most of the temporary skeleton of other vertebrate embryos and is the template on which most bones develop. It persists in adult mammals as parts of joints, in the respiratory passages and in the pinna of the ear.

SAQ 30 The cartilage end-plates of the intervertebral discs, and indeed the whole matter of the discs, contain collagen fibres (fibrocartilage). Suggest a reason for this.

SAQ 31 Elastic cartilage is found in the pinna of the ear and epiglottis. What property would you expect this type of cartilage to possess?

2.3.4 Bone

Bone is much harder than cartilage. The ground substance of bone is made up of 80% by mass of inorganic material, mainly calcium phosphate in the form of needle-shaped crystals of hydroxyapatite. The remaining 20% is organic material, mainly collagen fibres. This intercellular bone substance is produced by osteoblasts. Cells within the bone, called **osteocytes**, lie in cavities called **lacunae** within the ground substance. They are osteoblasts which have ceased to divide. Cytoplasmic processes from the cells connect with surrounding cells through tiny tubes in the ground substance, called **canaliculi**. Each osteocyte has direct or indirect connections with one of the many blood vessels which are found in bone. Osteoblasts and osteocytes are shown in figure 18.

There are two types of bone, spongy bone and compact bone.

Spongy bone

This is made up a network of apatite crystals (a compound of calcium) which are approximately 40 nm in diameter and are not visible to the naked eye. The collagen fibres point in all directions. Spongy bone is fluid-filled and less dense than compact bone (see figure 19(a)). It is usually found in the centre of the ends of long bone.

19 (*a*) **Spongy bone** × 200 (*b*) **Compact bone** × 100

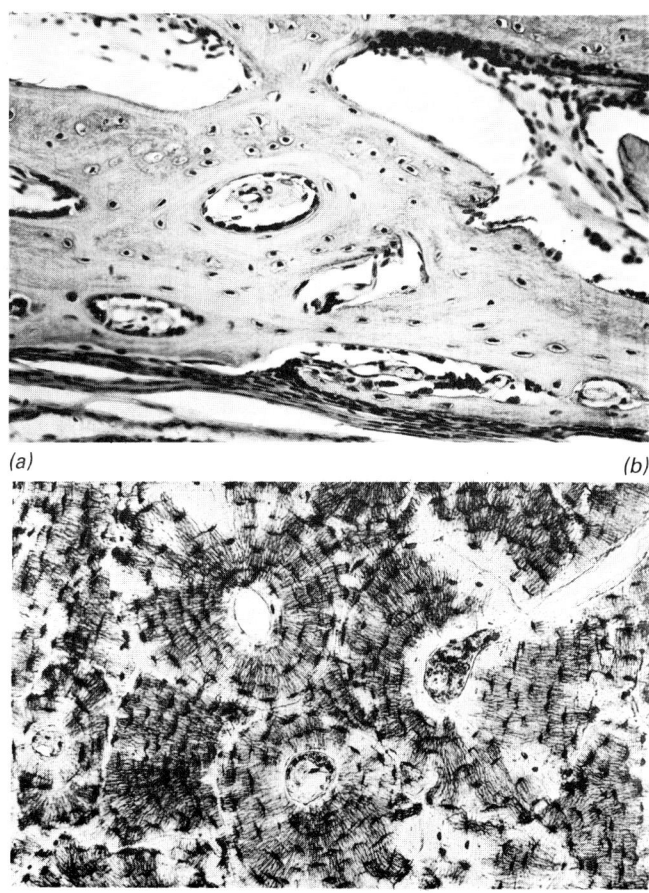

(a)
(b)

18 Bone cells

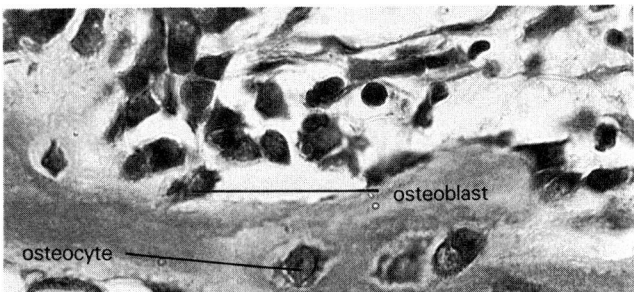

osteoblast

osteocyte

Compact bone

In this, the apatite crystals are very closely packed and are not easily distinguishable. The collagen fibres are usually arranged in sheets or lamellae. Cylindrical cavities are often eroded away around blood vessels in bone. New bone is subsequently laid down to fill the cavity as a series of concentric lamellae. These are known as **Haversian systems.** The central cavity, known as the Haversian canal, contains blood vessels and nerves. This is shown in figure 19(*b*) and figure 20.

20 Haversian systems

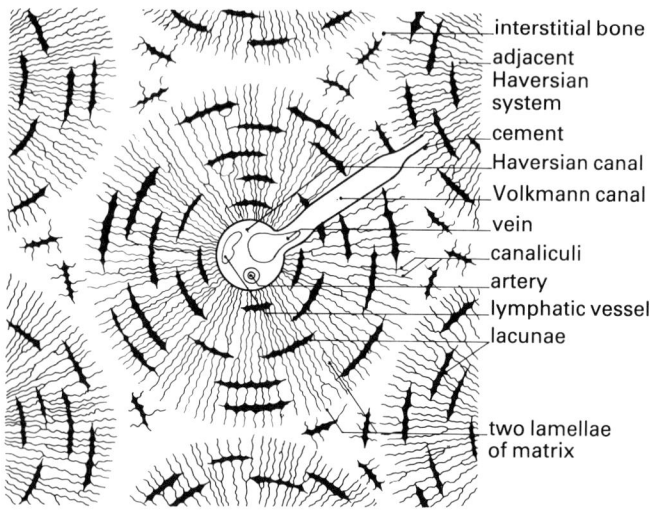

- interstitial bone
- adjacent Haversian system
- cement
- Haversian canal
- Volkmann canal
- vein
- canaliculi
- artery
- lymphatic vessel
- lacunae
- two lamellae of matrix

The collagen fibres toughen bone making it less brittle whilst, at the same time, they give a certain flexibility. The characteristic hardness of bone is due to the apatite crystals.

Bone substance is continuously recycled. In a human adult, about 0.05% of the total skeleton is replaced each day.

SAQ 32 Choose from the following words or expressions to match the statements (*a*)–(*d*). Collagen fibres, elastic fibres, bone, cartilage.

(*a*) The two tissues best at resisting compression.
(*b*) The hardest substance in the skeleton.
(*c*) The tissue with the greatest tensile strength.
(*d*) The tissue which can withstand the greatest temporary deformation.

2.3.5 Growth and development of bone

Bone can be formed in two different ways.

(*a*) Limb bones, the vertebrae and the girdles form from already existing cartilage. These are known as **cartilage bones.**
(*b*) Initially, parts of the skull and the clavicle are formed directly from connective tissue in the embryo. These are known as **membrane bones.**

Figure 21 shows the formation of cartilage bone.

21 Growth and development of long bone from cartilage

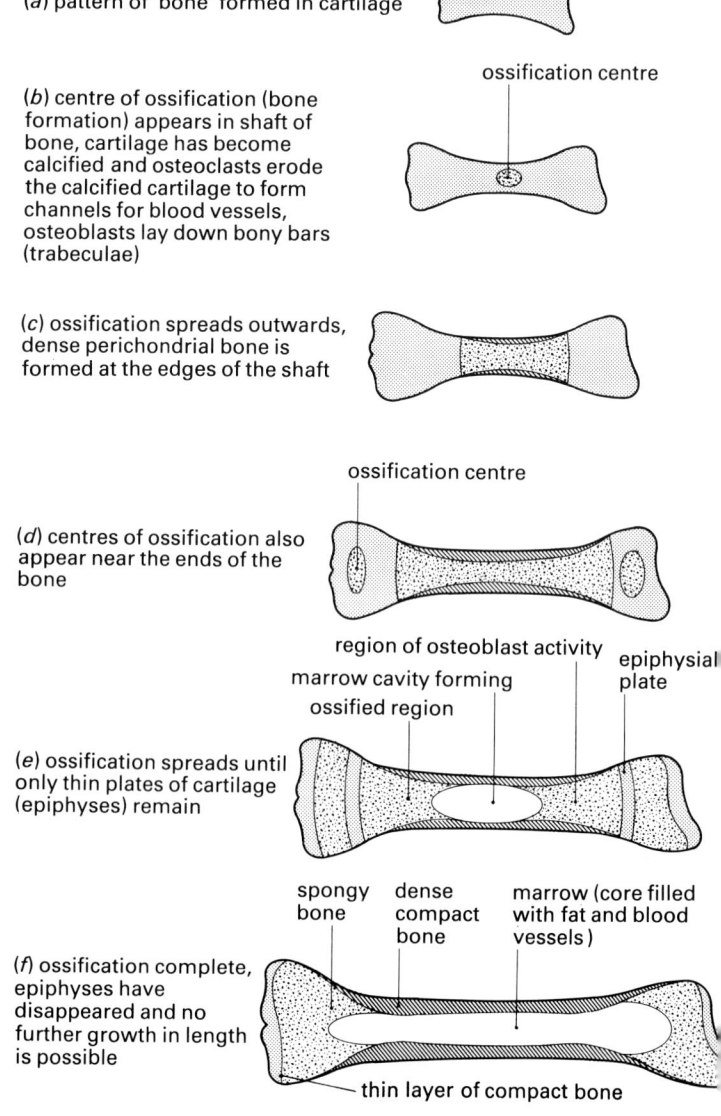

(*a*) pattern of 'bone' formed in cartilage

ossification centre

(*b*) centre of ossification (bone formation) appears in shaft of bone, cartilage has become calcified and osteoclasts erode the calcified cartilage to form channels for blood vessels, osteoblasts lay down bony bars (trabeculae)

(*c*) ossification spreads outwards, dense perichondrial bone is formed at the edges of the shaft

ossification centre

(*d*) centres of ossification also appear near the ends of the bone

region of osteoblast activity
epiphysial plate
marrow cavity forming
ossified region

(*e*) ossification spreads until only thin plates of cartilage (epiphyses) remain

spongy bone   dense compact bone   marrow (core filled with fat and blood vessels)

(*f*) ossification complete, epiphyses have disappeared and no further growth in length is possible

thin layer of compact bone

Membrane bones are formed from areas of connective tissue in the embryo. The **ossification** process is similar to that described in figure 21. Fibres are laid down. Calcium salts are deposited around the fibres. **Osteoclasts** erode the calcified matrix and, finally, **osteoblasts** form the bony **trabeculae**. Enlargement of skull bones occurs by addition of new bone externally and its removal from the inner side. This is necessary in order to increase the radius of curvature of the skull.

Using information from figure 21, answer the following questions.

SAQ 33 Compare the function of osteoblasts and osteoclasts.

SAQ 34 Where are the centres of ossification in a long bone?

SAQ 35 Answer (a)–(c) with reference to cartilage and membrane bones.
(a) What tissue is it derived from?
(b) What tissue is it made of?
(c) Give two examples of each type of bone.

2.4 Summary assignment 2

1 Make a copy of figure 14: skeletal materials based on collagen.

2 List other organic skeletal materials and state where each occurs.

3 Complete the table shown in figure 22. Part has been done for you.

4 Make brief notes on membrane and cartilage bones to explain (a) how they are formed, and (b) where they are formed.

2.5 Muscle

Muscle tissue is made up of elongated cells. These cells have the ability to contract. This enables them to exert a force on structures to which they are attached, such as bone, or within structures of which they are part, such as the walls of the intestine, arteries, the heart and the body itself.

There are three types of muscle which you will study. They are (a) striated muscle, (b) smooth muscle and (c) cardiac muscle.

Striated muscle

This is made up of many fibres. These may measure from 1 mm to more than 100 mm long and up to 100 μm in diameter. Each one is a multinucleate structure called a **syncytium**. The fibres under the light microscope show striations (figure 23); hence the term striated muscle.

Smooth muscle

This is composed of elongated cells containing a nucleus. They are smaller than those of striated muscle, being up to 0.5 mm in length and 6 μm in diameter (figure 24). Smooth muscle movements are not under voluntary control and this tissue is found in such regions as the walls of the alimentary canal, bladder, blood vessels and ducts of glands and the dermis of the skin.

22 **Summary of the structure and functions of connective tissue**

Tissue	Cells	Matrix	Fibres	Properties
areolar connective tissue				
fibrous connective tissue	fibroblast	mucopolysaccharides	collagen elastic	flexible, great tensile strength, resists deformation, can stretch and return to original length
cartilage				
bone				

23 Striated muscle fibres as seen under the light microscope × 2000

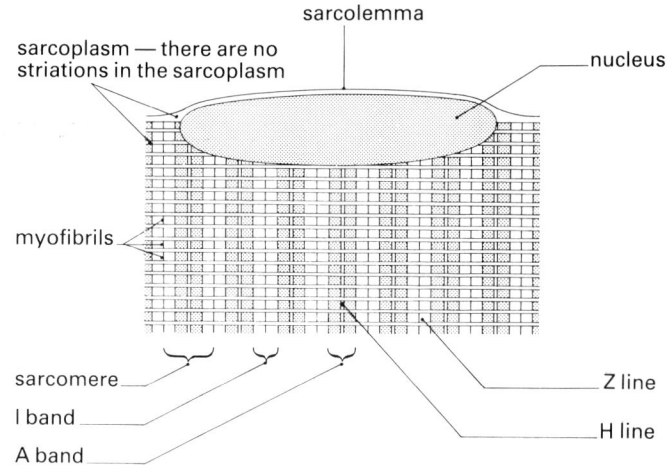

sarcolemma

sarcoplasm — there are no striations in the sarcoplasm

nucleus

myofibrils

sarcomere

I band

A band

Z line

H line

24 Smooth muscle fibres as seen under the light microscope × 800

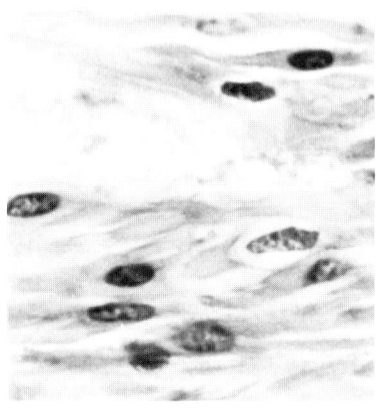

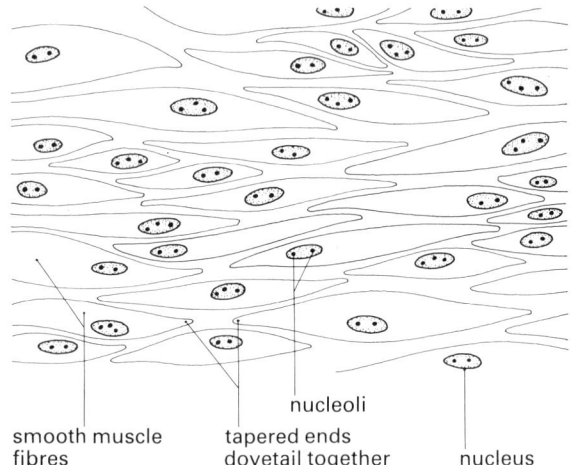

nucleoli

smooth muscle fibres

tapered ends dovetail together

nucleus

25 Cardiac muscle × 1100

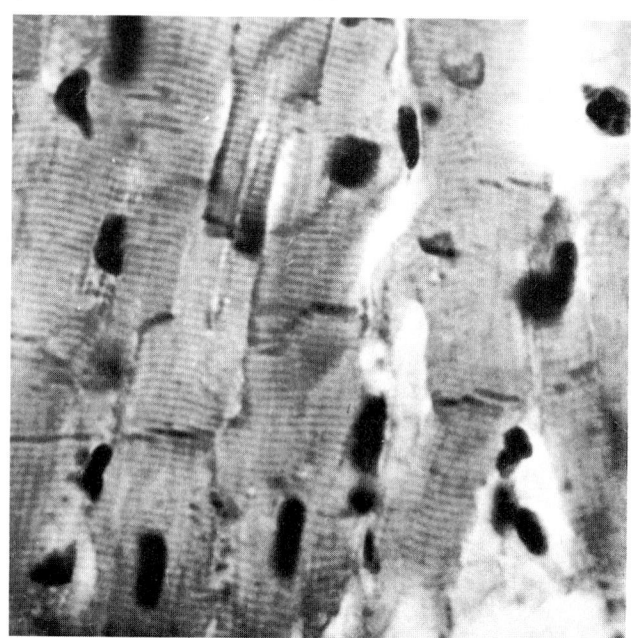

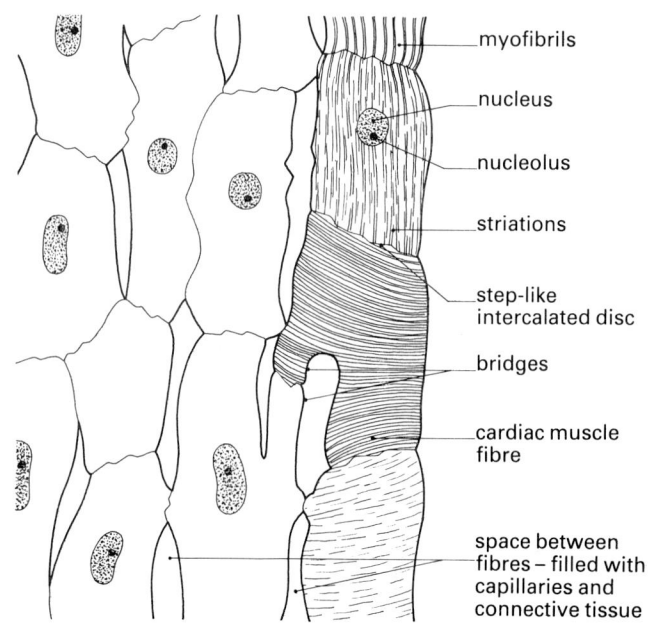

myofibrils

nucleus

nucleolus

striations

step-like intercalated disc

bridges

cardiac muscle fibre

space between fibres – filled with capillaries and connective tissue

Cardiac muscle

As its name suggests, this muscle is found in the heart. It is composed of fibres which are divided into short, cylindrical uninucleate cells by partitions known as **intercalated discs**. The cells are 4–5 μm long and 2–3 μm wide. The fibres are striated, branch and **anastomose** (join up with each other) to form an elaborate 3D net-like arrangement (figure 25).

The cytoplasm of muscle cells is called **sarcoplasm**. Each fibre of a striated or skeletal muscle is surrounded by a thin, connective tissue membrane, the **sarcolemma** or **endomysium**. Groups of fibres are bundled together within a connective tissue layer containing blood and lymph vessels and nerves. This is known as the **perimysium**. Finally, the whole muscle, made up of many such groups of fibres, is surrounded by the outermost connective tissue layer, the **epimysium**.

26   The internal structure of a skeletal muscle

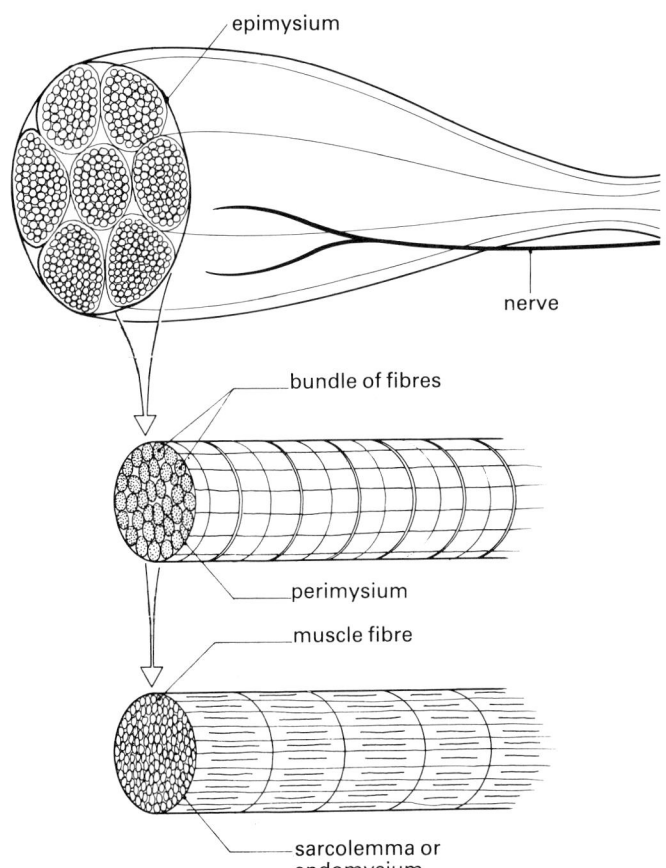

SAQ 36 Compare the size, shape and structure of a striated, a smooth and a cardiac muscle cell.

2.6 Mammalian support materials

In the practical which follows, you will examine connective tissue, bone, muscle and cartilage with the naked eye and under the microscope. You will also have the opportunity to make a temporary slide preparation of muscle tissue. Finally, you will test your ability to recognise these tissues in a slide of an embryo mammal.

Practical A: Examination of tissues involved in support

Materials

Long bone from the butcher with some meat still on sawed in half longitudinally, prepared slides of connective tissue with collagen and elastic fibres, cartilage, spongy and compact bone, striated, smooth and cardiac muscle, a prepared slide of TS through a young mammal in the region of the thorax, microscope and lamp

Procedure

(*a*) Examine the external features of the bone and the internal features revealed by the longitudinal section. Identify spongy and compact bone and cartilage (use figures 17, 19, 20 and 21 to help you), muscle (meat), and the structures which attach the muscle to the bone (tendons) which are 90% collagen.

(*b*) Make annotated drawings of your observations.

(*c*) Examine prepared slides of connective tissue with collagen and elastic fibres, cartilage, spongy and compact bone, smooth, striated and cardiac muscle. Use figures 15–25 to help you.

(*d*) Make annotated diagrams of your observations.

(*e*) Examine the prepared slide of the TS through the mammalian thorax. Identify the following features:

body wall, spinal cord, vertebral column, alimentary canal, heart, lungs, ribs and intercostal muscles (between the ribs).

(f) Now identify the major regions on the slide of the following tissues. Guidance as to where to find these tissues is given in brackets.

Striated muscle (body wall, associated with limbs and ribs)
Smooth muscle (alimentary canal walls)
Cardiac muscle (heart)
Cartilage (limbs)
Bone (limbs and ribs)

(g) Make a tissue map of the specimen and label the main features. Show the distribution of tissues listed in (f) by means of shading and a key.

Show this work to your tutor.

Extension practical B: Slide preparation of striated muscle

Materials

Mounted needle, fine forceps, 10 small watch glasses, 50 cm³ 70% alcohol, 10 cm³ Erhlich's haematoxylin, 10 cm³ acid alcohol, 10 cm³ alkaline alcohol, 10 cm³ 90% alcohol, 20 cm³ 100% alcohol (industrial methylated spirit), clove oil, Canada balsam, slides and cover-slips

Procedure

Using fresh muscle from the bone, make a temporary mount in the following way.

(a) Label the watch glasses 1–10. Place the reagents in the watch glasses as follows: 1, 4, 6, 70% alcohol; 2, Erhlich's haematoxylin; 3, acid alcohol; 5, alkaline alcohol; 7, 90% alcohol; 8 and 9, 100% alcohol; 10, clove oil. Steps (b)–(k) follow this sequence of watch glasses.

(b) Tease out the muscle fibres with mounted needles in 70% alcohol. Separate out short pieces of fibre without breaking.

(c) Stain in Erhlich's haematoxylin for 5 min.

(d) Cover in acid alcohol to differentiate, that is to remove the colour from most of the specimen, but leave it in the structures that hold it particularly strongly.

(e) Wash in 70% alcohol.

(f) Cover with alkaline alcohol to 'blue'.

(g) Wash in 70% alcohol.

(h) Cover with 90% alcohol for 1 min.

(i) Place in two changes of 100% alcohol for 5 min.

(j) Clear in clove oil for 5 min. Tease again.

(k) Mount in Canada balsam.

Examine your slide under the microscope and identify the fibres, striations, myofibrils and muscles. Make a drawing of your preparation. Add notes on how it was made.

Show this work to your tutor.

2.7 The ultrastructure of striated muscle

In your observations of muscle under the light microscope you saw that each fibre was made up of multinucleate fibres. The fibres have transverse stripes. You may also have seen longitudinal lines running along the fibres.

The electron microscope has revealed a much more detailed structure (figure 27). It is important to know about this in order to understand how muscle actually works to support animals and enable them to move.

The apparent longitudinal lines shown in an electron micrograph of muscle tissue indicates that each fibre is made up of a bundle of smaller fibres called **myofibrils**. These are embedded in the cytoplasm (sarcoplasm) of the fibre. Also in the sarcoplasm are mitochondria and a network of membrane-bound tubes and spaces. These features are shown in figure 28. They include a system of transverse channels opening at the surface of the sarcolemma, called the

T-system, and a closed system of channels and sacs called the **sarcoplasmic reticulum** (SR).

27 Electron micrograph of a skeletal muscle × 140 00

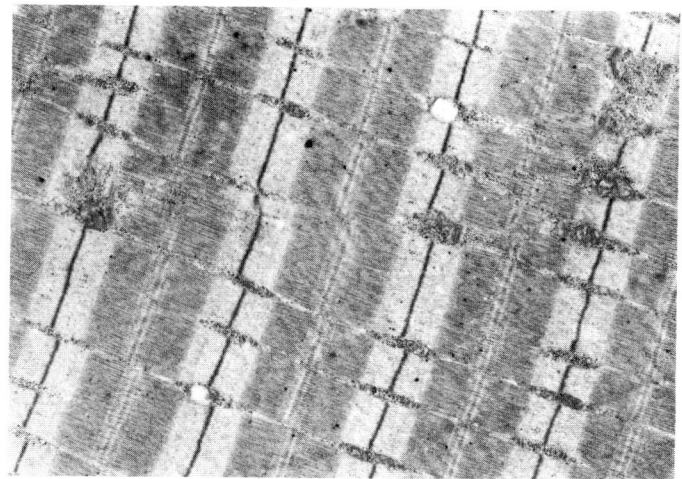

28 Part of a muscle fibre

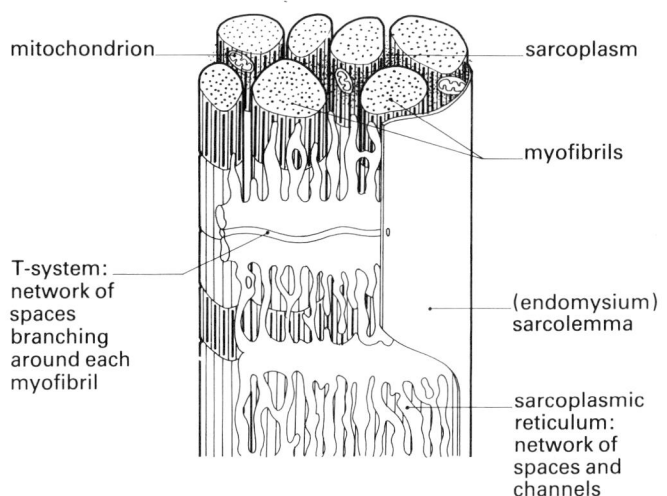

mitochondrion

sarcoplasm

myofibrils

T-system: network of spaces branching around each myofibril

(endomysium) sarcolemma

sarcoplasmic reticulum: network of spaces and channels

The transverse striations are caused by the arrangement of two types of filament within the myofibril. Each filament is made up of bundles of protein fibres. There are the thick filaments of the protein **myosin** and the thin filaments of the protein **actin**. In addition to the helical actin molecules, the thin filaments of striated muscle also contain the regulatory proteins **tropomyosin** and **troponin**. These structural details have been worked out using the electron microscope and X-ray diffraction.

29 Arrangement of protein filaments in a sarcomere of a myofibril

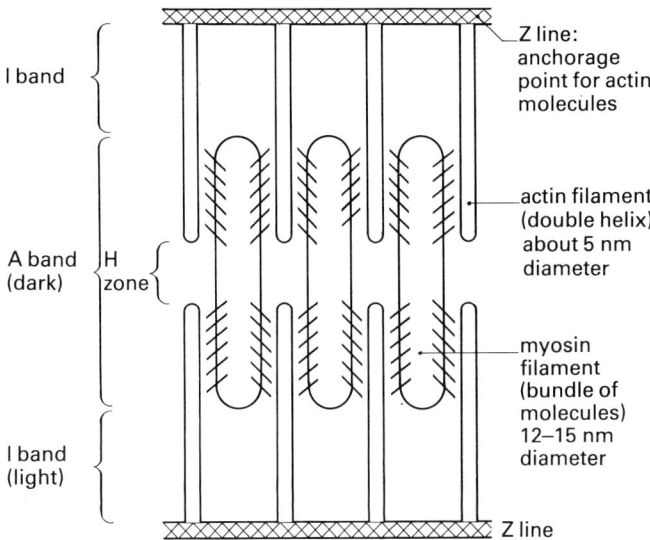

I band

A band (dark)   H zone

I band (light)

Z line: anchorage point for actin molecules

actin filament (double helix) about 5 nm diameter

myosin filament (bundle of molecules) 12–15 nm diameter

Z line

Using information from figure 29, answer the question below.

SAQ 37 What types of fibre are found in the following regions?
(a) A band (b) I band (c) H zone

Myosin molecules are rod-shaped with a globular end, like a double-headed golf club (see figure 30(a)). Figure 30(b) shows one possible arrangement of the molecules of actin and myosin filaments. Figure 30 (c) shows the pattern of arrangement of actin and myosin filaments in myofibrils, based on an EM of a cross-section of myofibrils.

30 Myosin molecules and the possible pattern of their arrangement in a myofibril: (a) electron micrograph of myosin molecules × 75 000;

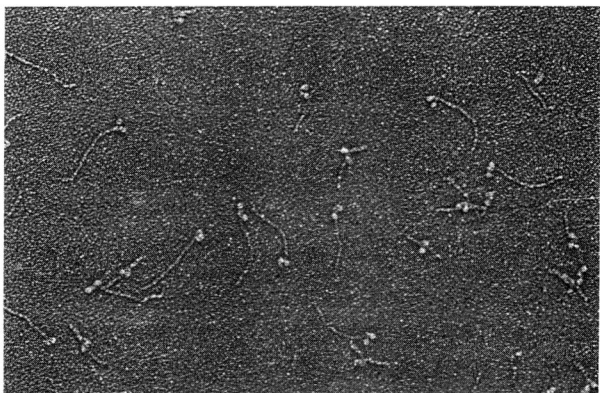

(b) possible arrangement of actin and myosin molecules;

(b)

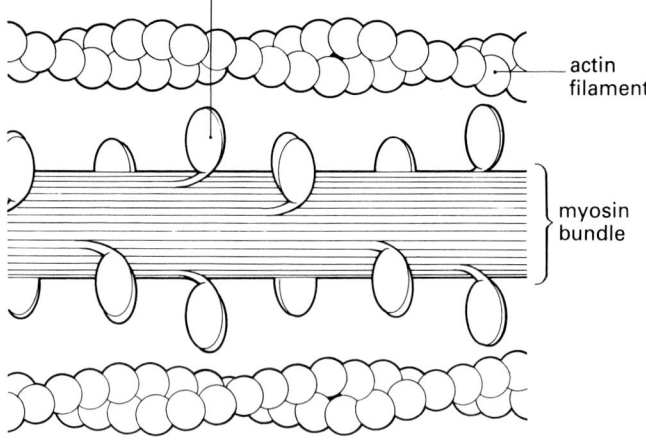

heads move towards and bind with globular subunits of actin to form cross-bridges

actin filament

myosin bundle

(c) cross-section through a myofibril

(c)

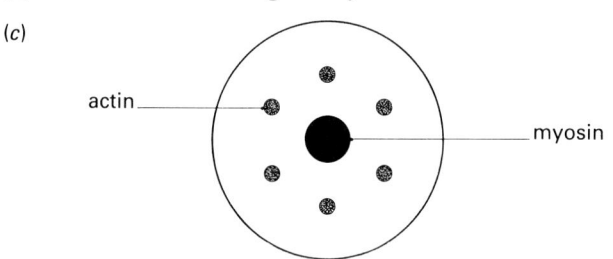

actin

myosin

2.8 The chemistry and energetics of muscle contraction

Figure 31 shows the changes that can be observed with the electron microscope in muscle myofibrils during contraction. Study the diagrams and then answer the questions below.

SAQ 38 During contraction, what happens to (a) the A band, (b) the I band, (c) the sarcomere length, (d) the lengths of the actin and myosin filaments?

It is thought that during contraction the actin and myosin filaments slide over each other. This must involve the breaking and remaking of the cross-bridges between the two types of molecule. Contractile force is generated by the ratchet-like detachment and attachment of the myosin heads. The energy required for this comes from the breakdown of ATP. Myosin molecules can act to catalyse the breakdown of ATP. Their catalytic ability is activated when they are bound to actin. It

31   Changes in myofibrils during contraction

(a)

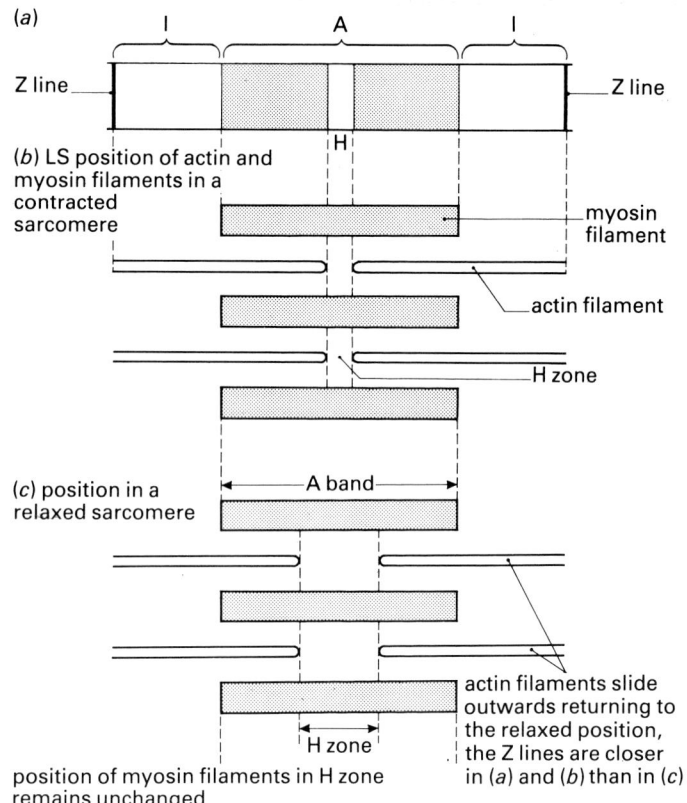

Z line

Z line

H

(b) LS position of actin and myosin filaments in a contracted sarcomere

myosin filament

actin filament

H zone

(c) position in a relaxed sarcomere

A band

H zone

actin filaments slide outwards returning to the relaxed position, the Z lines are closer in (a) and (b) than in (c)

position of myosin filaments in H zone remains unchanged

is inhibited by the molecules troponin and tropomyosin and calcium ions are involved in this process.

Because there are only small stores of ATP in striated muscle fibres, continued contraction depends upon continuous synthesis by the mitochondria. If the ATP level becomes limiting, a muscle fibre remains permanently contracted and is said to be in rigor (as in *rigor mortis*).

2.8.1 Nervous stimulation of muscle fibre contraction

Contraction is initiated by a nerve impulse. The relationship between a nerve fibre and a muscle fibre is illustrated in figures 32 and 33. Notice that the membranes of the two types of fibre lie close together but do not fuse or touch. The region of association is known as the **motor end-plate**. Neuromuscular junctions are also studied in section 4.5 of the unit *Response to the environment*.

32 Motor end-plate. V shows some neuromuscular junctions

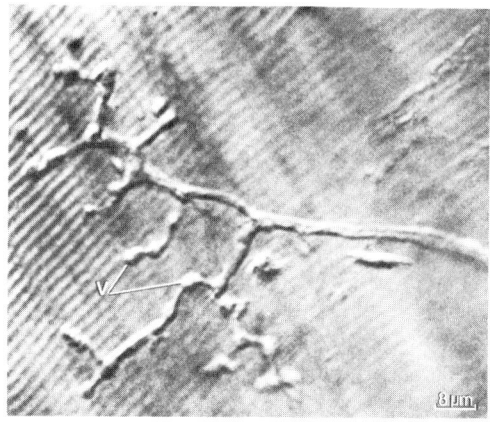

33 Diagram to show the structure of a neuromuscular junction

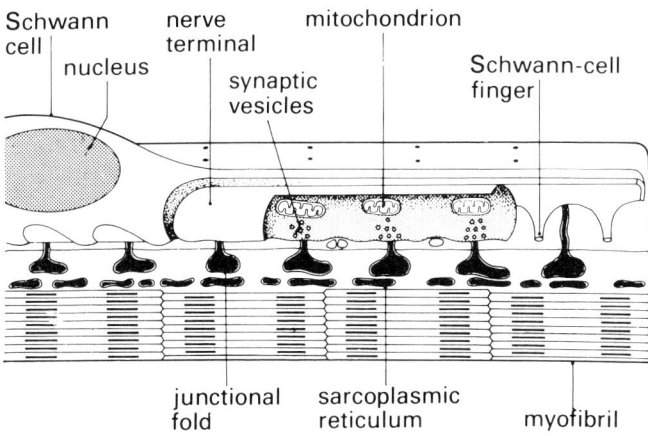

The terminal branches of the axon contain many synaptic vesicles which contain the chemical transmitter acetylcholine (ACh). When an action potential arrives at the end of the nerve fibre, acetylcholine is released into the space between the axon terminal and muscle membrane. The sarcolemma has a resting potential of 50–100 mV, with the inside being negative. ACh reacts with receptor molecules on the sarcolemma. This causes the latter to become highly permeable to small cations. Sodium ions flow in and the membrane is depolarised. Thus, an action potential is set up. The action potential is 30–40 mV, with the inside being positive. The action potential spreads over the surface of the muscle fibre, through the T-system and through the sarcoplasmic reticulum. It stimulates the release of Ca^{2+} from the tubules into the sarcoplasm. This removes the inhibitory action of troponin and tropomyosin. Hence, myosin is able to catalyse the breakdown of ATP and energy is released.

The effects of nervous impulses on whole muscles is studied in section 3.8 of this unit.

2.9 Turgor pressure and support in a young shoot

In young plants and all herbaceous plants, turgor pressure, in which cell sap serves as a kind of hydrostatic skeleton, is an important means of support. The cells under such pressure are mainly in the parenchyma tissue.

If two similar young, healthy, leafy shoots are taken and one (cut under water to prevent air entering the stem) is placed in a beaker of water and the other in an empty beaker, after a few hours the first shoot remains firm while the second begins to droop or wilt. Wilting occurs because water is lost from transpiration and is not replaced by an equivalent water uptake. This leads to a decrease in the water content of the vacuoles of the parenchyma cells. With less water there is no outward pressure on the cell walls so the turgor pressure of the cells fails and they become **flaccid** (less firm).

Intracellular support provided by turgor pressure within cells can be compared to the support provided by air in an inflated air-bed.

In Practical C it is assumed that you know about the following:

the structure of parenchyma cells, osmosis, water potential, osmotic potential, turgor, differentially permeable membranes.

If you feel unsure about any of these, go back and revise them from *Cells and the origin of life*, section 6.7, and *The continuity of life*, section 7.5.

Practical C: Turgor pressure and support in plant tissue

In this practical you will see how a change in the water content of parenchyma cells affects their ability to provide support.

Materials

Potato, cork borer, ruler (in mm), 3 pins, corked specimen tube about 6 cm tall filled with water, graph paper 7 × 5 cm, 3 corked specimen tubes about 6 cm tall, 15 cm³ sodium chloride solution, 1 mol dm⁻³ NaCl, 15 cm³ sodium chloride solution, 0.5 mol dm⁻³ NaCl, distilled water, sticky labels, cutting board

Procedure

(a) Label three specimen tubes **A**, **B** and **C**. Fill **A** with 1 M NaCl solution. Fill **B** with 0.5 M NaCl solution. Fill **C** with distilled water.

(b) Cut three identical strips of potato approximately 40 × 3 × 3 mm. The exact size is not important but all three should be the same (use a cork borer for this).

(c) Place one strip in **A** so it is completely covered. At 5 min intervals, place a strip in **B** and **C**.

(d) Leave each strip in its solution for 15 min.

34 Apparatus for practical C

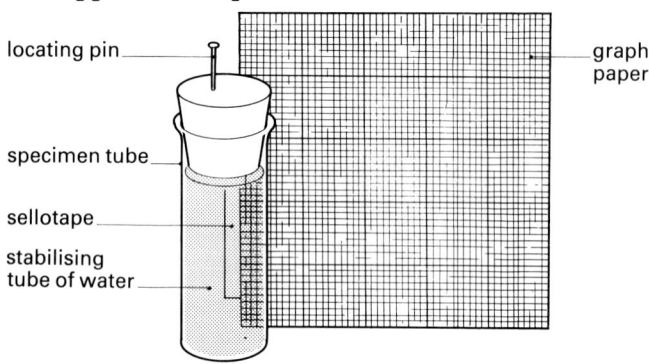

locating pin — graph paper

specimen tube

sellotape

stabilising tube of water

(e) Fix the graph paper to the specimen tube with sellotape and position the locating pin, as shown in figure 34.

(f) After each potato strip has been in its solution for 15 min, remove it from the tube. Notice how it feels. Push a marker pin in one end and then fix it to the cork of the specimen tube with a second pin, as shown in figure 35.

35 Investigating the strips

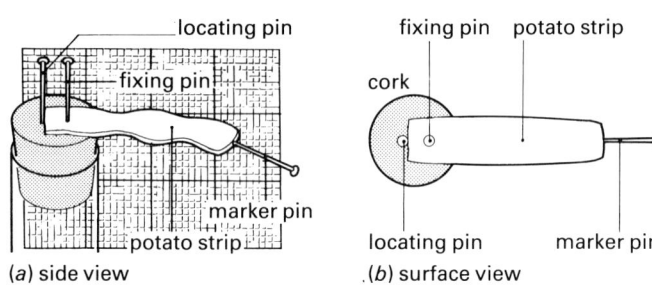

(a) side view   (b) surface view

(g) Devise a method for comparing the turgidity of the three potato strips.

(h) Record your results in a suitable form.

Discussion of results

1 How did the three potato strips feel on removal from the tubes?

2 What happened to the three strips when they were fixed to the cork?

3 Explain your observations. Refer to the following terms in your explanation: osmosis, water potential, osmotic potential, turgor, differentially permeable membrane, parenchyma cells.

4 Relate your findings from this investigation to the method of support in non-woody plants and plant parts.

5 Suggest ways in which the experimental techniques used here could be improved.

Show this work to your tutor.

2.10 Plant cells specialised for support

As plants increase in size, turgor pressure alone is insufficient to provide total support.

Two types of specialised support tissues are found in plants. These are collenchyma and sclerenchyma fibres.

Collenchyma is often found in a cylinder just inside the epidermis of stems, as shown in figure 37. The cells of this tissue are living but they have their walls thickened with extra layers of cellulose. These extra layers are often laid down unevenly. Thus, collenchyma can be recognised best by looking for cells with irregularly thickened walls. Figure 36 shows photographs of collenchyma. Study them carefully so you will be able to recognise collenchyma in a stem section.

36 Structure of collenchyma cells (*a*) LS × 200, (*b*) TS × 300

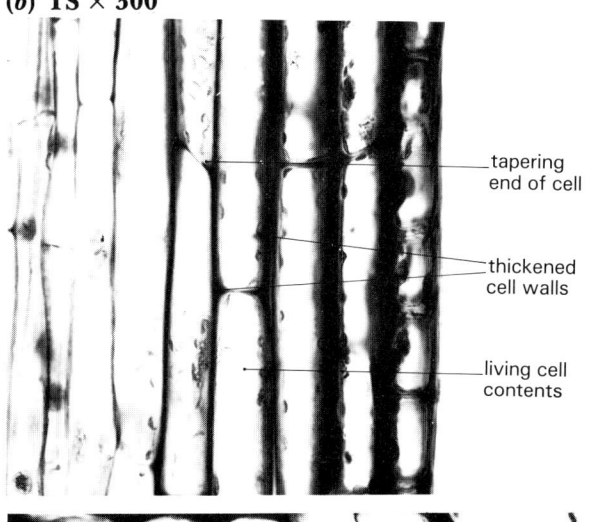

tapering end of cell

thickened cell walls

living cell contents

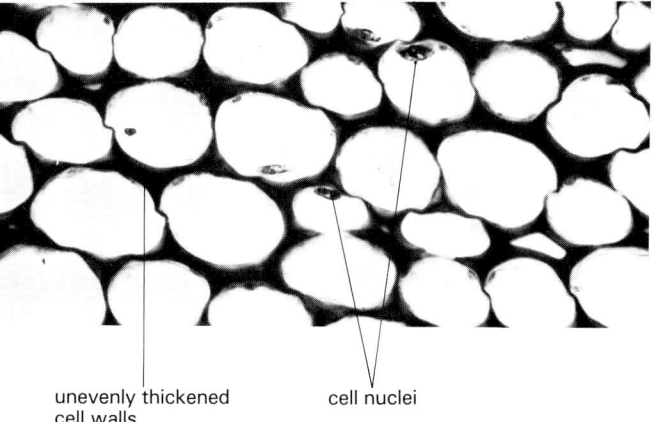

unevenly thickened cell walls   cell nuclei

SAQ 39 From your observations of figure 36, describe the three-dimensional shape of the collenchyma cell.

It is the lignified cells which are most effective in providing support in the larger plants. Although fibres are the most important lignified cells involved in the support of plants, there are other lignified cells, including the transporting cells of the xylem (vessel elements and the tracheids).

Sclerenchyma fibres may be found in cylinders within the cortex of stems and roots. This is more common in monocotyledons than dicotyledons (figure 37). Sclerenchyma is found associated with vascular bundles, within the bundle itself especially in the xylem and also just outside it (figure 37).

37 Distribution of collenchyma and sclerenchyma in plant cells: (*a*) TS portion of a young dicotyledon stem; (*b*) TS portion of a young monocotyledon stem

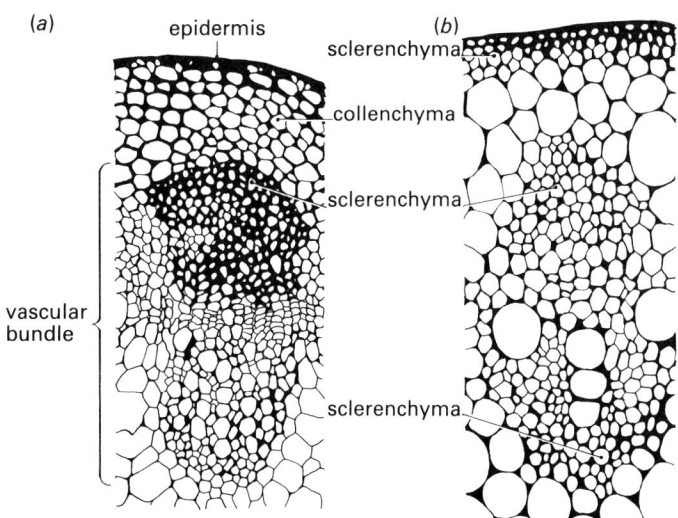

(a) epidermis   (b)
sclerenchyma
collenchyma
sclerenchyma
vascular bundle
sclerenchyma

Sclerenchyma fibres begin as living cells, but by the time they reach maturity and are fully functional they are dead. They have an elongated, roughly cylindrical shape. The walls become greatly thickened with a secondary layer of lignin which is laid down inside the primary cell wall. The inside of the cell is usually empty but may contain the remains of living cytoplasm. Figure 38 shows photographs of sclerenchyma. Study them carefully so you will be able to recognise sclerenchyma in a stem section.

38 Structure of sclerenchyma fibres (a) LS × 540, (b) TS × 450

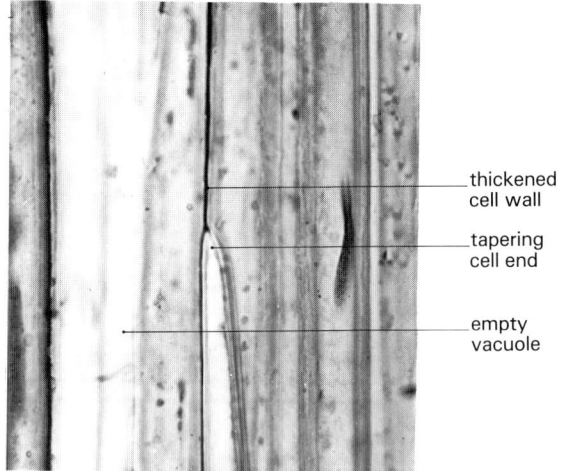

thickened
cell wall

tapering
cell end

empty
vacuole

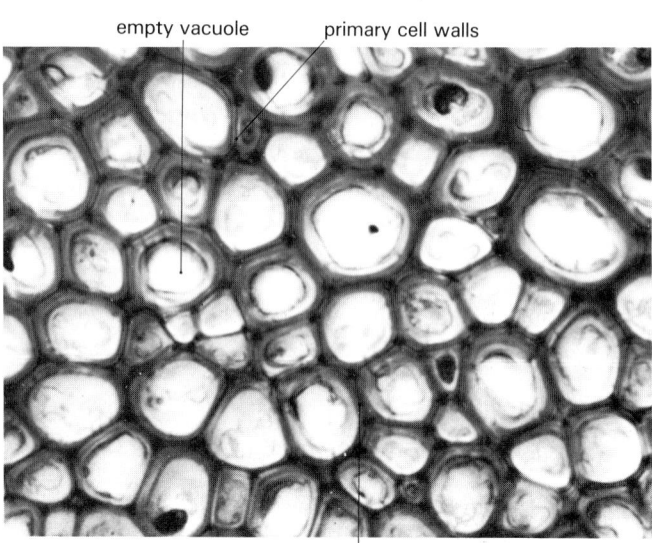

empty vacuole   primary cell walls

thick lignin layer

SAQ 40 List the similarities and differences between collenchyma cells and sclerenchyma fibres.

Two previous units have studied stem sections. *Continuity of life* 7.5 and *Exchange and transport* 6.3. Practical S in *Exchange and transport* looked particularly at lignified tissue in stem sections and macerates. Look back at your practical record and relate your findings in Practical S to the problem of support. If you did not carry out this practical, consult your tutor who may suggest supplementary work on stem sections.

Extension practical D: Forensic investigations and plant tissues

Forensic work often involves the identification of plant tissues which may link a suspect with a crime. The plant tissues involved may have been picked up from plants themselves. Alternatively, it may be necessary to examine man-made products based on plant tissues, such as paper.

There are two methods of preparing wood-pulp for paper manufacture: one is mechanical, the other is chemical. The chemical process removes much of the lignin.

In this practical you will macerate different types of paper and then observe the macerate under the microscope to identify the cells present and the nature of their walls.

Materials

Paper samples, such as newsprint from a 'popular' and a 'quality' newspaper, strawboard (the kind of cardboard used for backing student refill pads), typing paper, cigarette paper, etc., electric food-blender or beaker of water, bunsen burner, tripod, gauze and glass rod, 5 cm³ phenylammonium (aniline) chloride or sulphate in small bottle with dropper pipette, slides and cover-slips, mounted needle, microscope and lamp

Procedure

(*a*) Macerate your paper samples (small pieces only are required) either by swirling with plenty of water in an electric blender or by boiling in water.

(*b*) Place a few drops of phenylammonium stain on a clean slide. Pipette some of the macerate onto the drops. Mix with the mounted needle. Add a cover-slip.

(*c*) Examine under the high power of the microscope. Look for sclerenchyma fibres. Notice their shape and their staining properties.

(*d*) What other cell types are present?

(e) Make a labelled drawing to show the three-dimensional structure of sclerenchyma fibres. Note the type of paper manufacture used for each sample. Annotate your diagram to explain about this.

Show this work to your tutor.

_____

2.10.1 The effect of the environment on plant support tissues

Water has a higher specific density than air. This means that the former provides greater support for organisms (buoyancy). Most higher plants are terrestrial and have well-developed xylem tissues for support.

Figure 39 shows a section through the stem of an aquatic angiosperm.

39 Cross-section through a stem of an aquatic angiosperm (*Hippuris* sp., mare's tail)

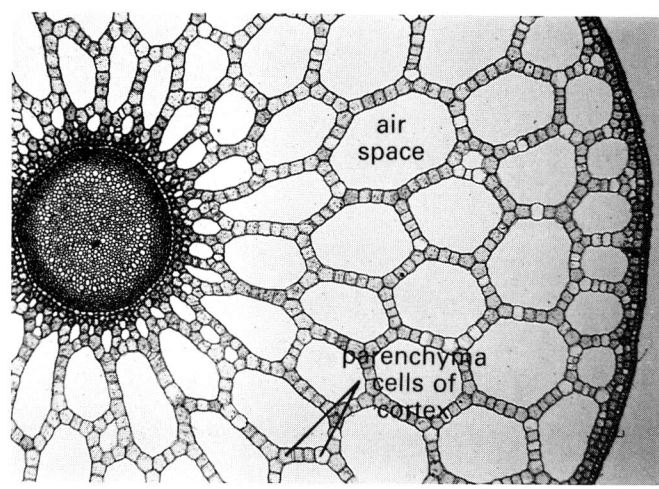

SAQ 41 What do you notice about the amount of xylem present in *Hippuris*?

The most striking feature of the *Hippuris* stem is the extensive system of air spaces. These spaces all interconnect to enable diffusion down to and up from the submerged parts of the plant. Such specialised tissue is common in aquatic plants and is known as **aerenchyma**.

Lower plants such as algae do not have xylem. They are mostly aquatic. Figure 40 shows the seaweed *Ulva lactuca* in and out of water.

SAQ 42 Describe the appearance of *Ulva lactuca* in and out of water.

SAQ 43 Suggest an important biological consequence of the effect of water on *Ulva lactuca* as shown in figure 40.

40   Seaweed (*Ulva lactuca*) (a) in and (b) out of water

(a)

(b)

2.11 The insect cuticle

The Arthropoda as a group of animals have, as one of their major characteristics, a hard external skeleton (exoskeleton). In insects, this exoskeleton, or cuticle, is secreted by the outermost layers of living cells.

The cuticle covers the whole of the body surface, but its thickness and composition vary. Thick, protective areas are linked by thinner regions known as **arthrodial membranes**. The latter give the exoskeleton, as a whole, considerable flexibility and permit movements to be made. The cuticle also lines the respiratory system, fore- and hindguts.

The cuticle comprises two main layers: an outer, thin **epicuticle** and an inner **procuticle**. Figure 41 shows the arrangement of layers in the insect cuticle.

41 Section through an insect cuticle

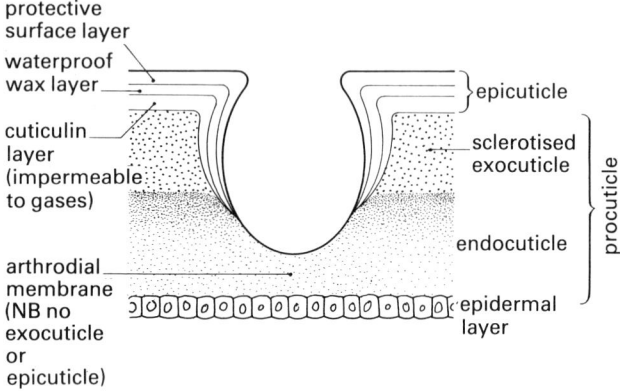

The epicuticle is seldom more than one micron thick and its layers lack chitin. It functions as a permeability barrier between the insect and its environment. The underlying procuticle may be differentiated into a sclerotised **exocuticle** and a softer **endocuticle**. It is composed of chitin fibres in a protein matrix. Chitin is a polysaccharide. The chitin fibres are arranged in layers. The orientation of fibres in each layer is different and this gives it strength in more than one direction, just like plywood.

The combination of chitin and protein results in a skeletal material which is very effective in providing support and protection. The chitin is stiff and resists tension. The protein withstands tension and is pliant. Chitin alone and in thin fibrils would buckle easily under compression, but the cross-linkages between the protein molecules reduce the likelihood of this.

Different chemical pathways are involved in the hardening process. Some of the complex cross-linkages include tyrosine derivatives, others involve tanning and the quinone links give rise to brownish cuticles. With the formation of cross-linkages, the fibrous chains become closely packed and this may result in dehydration and further hardening.

The cuticle also provides a firm surface for attachment of muscles (figure 42).

42 Relationship between cuticle and muscles

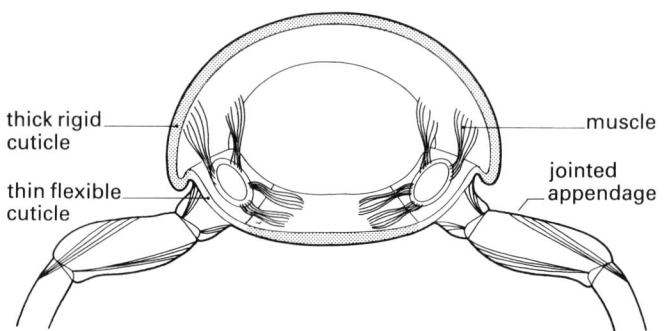

2.11.1 The insect cuticle and moulting

As an insect grows, the cuticle must be periodically shed since it is non-living itself and is therefore incapable of growth. Prior to the shedding of the cuticle, known as **moulting** or **ecdysis**, certain cells in the epidermis secrete a moulting fluid. This dissolves away the inner layers of the exoskeleton. At the same time, the epidermis increases its surface area by cell multiplication and begins to produce a new enlarged exoskeleton within the old remains.

Eventually, the old exoskeleton splits and the insect emerges with its new exoskeleton. At first, this is soft and so can stretch to allow for an increase in size. The insect may take in air or water to cause such an expansion. After a few hours, however, the new exoskeleton hardens around the enlarged insect. This process is controlled by hormones and is dealt with in unit 5, *Continuity of life*.

2.12 The advantages and disadvantages of an exoskeleton

An exoskeleton provides an excellent form of protection and support. It is also adaptable to allow flexibility and movement. Its major disadvantage is its inability to grow, resulting in the need for moulting.

SAQ 44 Suggest two problems an insect faces during actual moulting and while the new exoskeleton is hardening.

The second restriction imposed by the exoskeleton is on the size to which insects can grow. Despite their success as a group, none is very large.

As an animal increases in size, its volume increases by the power of three while its surface area increases by the power of two. Its mass increases with its volume while the exoskeleton increases with the surface area, assuming its thickness remains constant.

SAQ 45 As an insect increases in size, how will the increase in mass compare with the increase in exoskeleton?

To support itself, a larger insect would need a much thicker exoskeleton. This would greatly reduce the mobility and flexibility of the animal. Also, a thicker exoskeleton would itself increase the mass.

2.13 Summary assignment 3

1 Make diagrams to show the structure of the three types of muscle. Annotate to include information from your answer to SAQ 36.

2 Draw diagrams based on figures 27–31. Annotate these diagrams to explain the ultrastructure of muscle and its functioning.

3 Draw up a table to summarise the mechanical tissues of plants and their functions. Include information under the following headings: distribution of tissue, special adaptations, chemical nature of material giving support.

4 Make outline notes on the structure and properties of the cuticle of insects.

5 Note the advantages and disadvantages of exoskeletons.

Self test 2, page 110, covers section 2 of this unit.

2.14 Past examination questions

1(*a*) Give an illustrated account of the structure of the various types of mechanical tissue in a *named* herbaceous angiosperm.　　　　(10 marks)
(*b*) Briefly describe the distribution of these tissues in this plant.　　　　(6 marks)
(*c*) What other roles may be played by these tissues.　　　　(4 marks)

(University of London, 1980)

2 Structure and function are closely related. By reference to (*a*) striated (skeletal) muscular tissue, and (*b*) parenchyma in plants, discuss how far this statement is true.

(University of London, 1983)

Section 3   Forces, skeletons and joints

3.1 Introduction and objectives

A force may be defined as a measurable and determinable influence tending to cause the *movement* of a body. All organisms are subjected to forces acting upon their bodies from the environment. In turn, their bodies exert a force on the environment. When considering the support and movement of living things it is useful to be reminded of important points from Newton's three laws of motion. These may help you to understand the implications of forces.

1 A body remains at rest or moves at a constant velocity unless forces act upon it.

2 An unbalanced force gives a body an acceleration in the direction of the force.

3 If body **A** exerts a force on body **B**, body **B** exerts an equal and opposite force on body **A**.

This section considers the various forces acting upon organisms and the adaptations of organisms for stability and controlled movement. The main emphasis is on vertebrates and on human beings in particular.

After completing this section you should be able to do the following.

(*a*) State the four main forces acting on the bodies of animals.

(*b*) Describe the role of the following in the support system of a mammal: bones, muscles, nerves, limb arrangement, braces, counterbalancing cantilevers, arches, centre of gravity.

(*c*) Recognise and name the main bones of the human skeleton.

(*d*) Give the plan of a pentadactyl limb and describe its adaptations for swimming, flying, running, digging and so on.

(*e*) List the variety of joints found in the vertebrate skeleton and explain how they are adapted for their special functions.

(*f*) Recognise and interpret kymograph traces of muscle action in response to single and multiple stimuli.

(*g*) Explain the terms twitch, summation, tetanus and refractory period.

(*h*) Describe overall control of skeletal muscle systems.

(*i*) State the importance of tendons and ligaments.

(*j*) Explain the action of levers.

(*k*) Describe movement in mammals in terms of lever action.

(*l*) Name the main forces acting upon trees.

3.2 Forces acting on the body

Figure 43 illustrates situations in which various forces are acting upon the bodies of animals or are being exerted by animals. These forces can be classified into four main types.

In each of the diagrams in figure 43 the arrows represent the direction of the forces and the photographs show examples of the forces in action.

SAQ 46 Which of the forces shown in figure 43 will be important in the following situations?
(*a*) A browsing animal stretching up to eat leaves immediately in front of it.
(*b*) A dolphin swimming in rough seas.
(*c*) Young puppies playing and mock fighting.

The body is adapted in various ways to counteract these forces, as outlined in sections 3.3 and 3.4.

43 Types of force

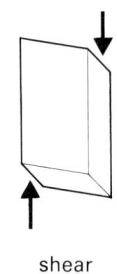

shear

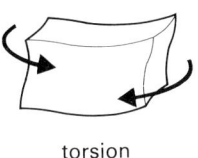

torsion

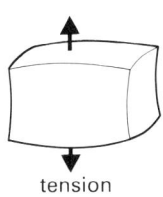

tension

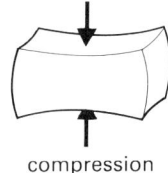

compression

44 Forces in action

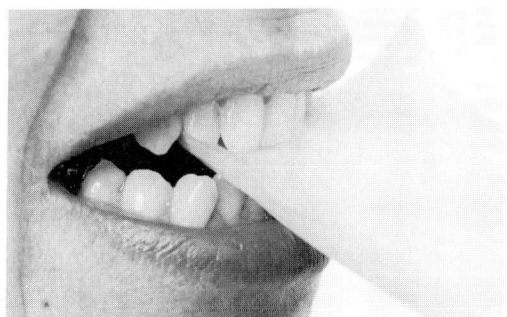

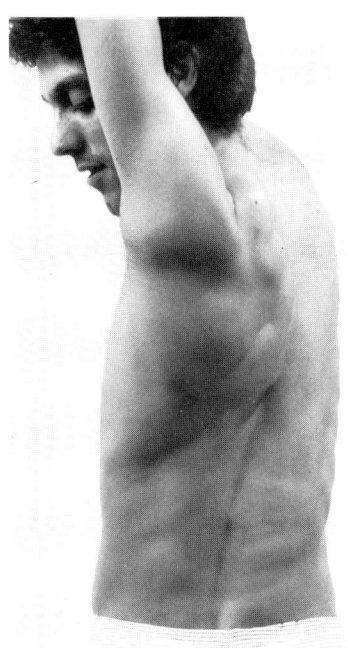

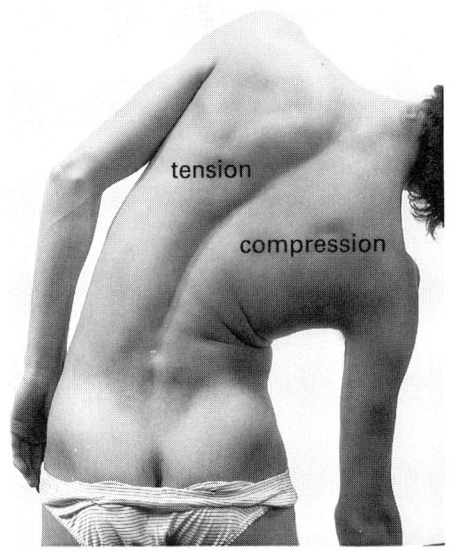

tension

compression

3.3 Stability and the centre of gravity

The weight of the body can be thought of as concentrated at a point in the body called the **centre of gravity**. In humans, it is estimated that this point is about 50 mm below the navel and just in front of the vertebral column. The reason for this estimate is that it changes with each beat of the heart, every breath taken and with the ingestion of food. Variations of 0.5–0.6 of a centimetre have been reported from forced inspiration.

If the body is tilted so the centre of gravity lies outside the base of the body, the body becomes unstable and will fall unless action is taken (figure 45).

45   Change in position of the centre of gravity leads to instability

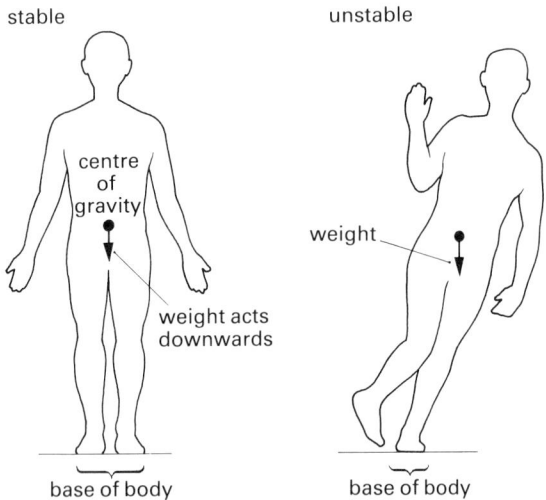

The position of the centre of gravity affects the stability of the animal at rest as well as affecting the kinds of movements it can make during locomotion. You will investigate this idea further in practical E. Remember the importance of the inner ear receptors in maintaining stability in balance. Their role is studied in unit 6, *Response to the environment*.

Practical E: Investigations on the length and position of legs in relation to stability

In this practical you will use models to investigate the effects of length and position of legs on stability of the body.

Materials

Plasticine block 2×1×1 cm, 3 straws, scissors, protractor, cardboard shoe-box

Procedure

(*a*) Make a tilting apparatus, as shown in figure 46.

46   Tilting apparatus

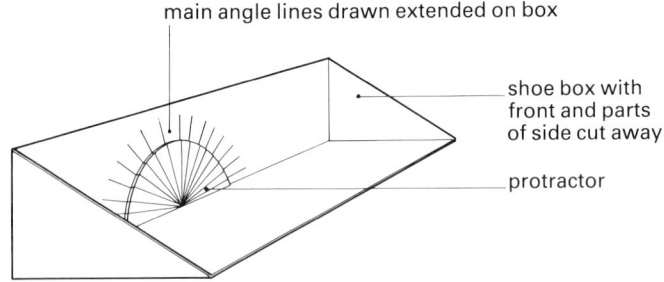

(*b*) Make a model animal, as shown in figure 47.

47   Model animal

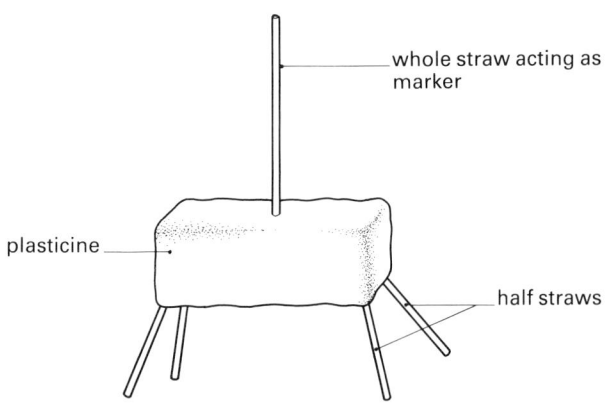

(*c*) Place the animal in the centre of the tilting apparatus with the marker in line with the 90° mark and slowly tilt the model sideways. Note the angle at which the model falls over, using the whole straw as a marker.

(*d*) Using this technique, design experiments to test the effects of leg length and the angle at which legs are inserted in the body on the stability of the model.

(*e*) Design a test to determine the position of the centre of gravity in each model you make.

(*f*) Write a full report of your experimental methods and present your results in a suitable form.

Discussion of results

1 Suggest ways in which the experimental techniques could be improved.

2 Make a general statement on the effect of leg length and centre of gravity on stability.

3 What factors would you expect to affect (*a*) the upper limit, and (*b*) the lower limit, of leg length?

4 Comment on what you would expect to be the relative leg length to body size in relation to locomotion in (*a*) tree-dwelling mammals, such as squirrels, and (*b*) in mammals of open plains, such as gazelles and zebra. Explain your answer.

5 How does the angle of the legs to the body affect stability?

6 What other factors, besides stability, will be affected by the angle of the legs to the body?

Show this work to your tutor.

3.4 Supporting body weight

Mammals have paired limbs which, in most terrestrial forms, are used as struts to support the body off the ground. They also play an important role in locomotion, as will be seen later.

The series of diagrams in figure 48 shows how the vertebrate limbs are thought to have evolved from the fins of fish into structures efficiently adapted for support on land.

SAQ 47 Why does a lizard need a relatively larger pelvic girdle and adductor muscles (which pull the limbs inwards towards the body) than a mouse?

Practical F: Investigations on the strength of legs in relation to body weight

In this investigation you will use models to see how variation in leg dimensions affects the body weight which can be supported.

48 Evolution of vertebrate limbs

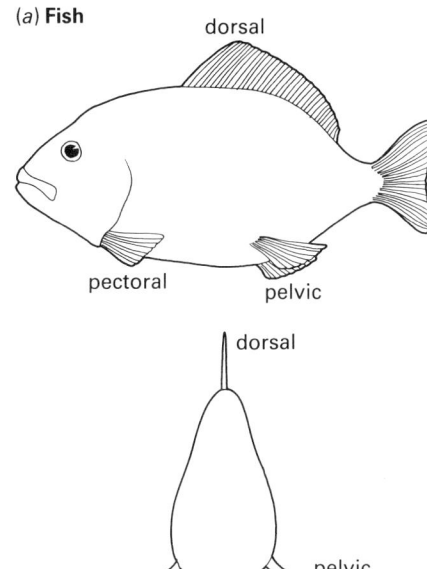

(*a*) **Fish**

Fins have no role in support except for a few oddities, such as the mudskippers, which use them as props. The paired limbs of primitive fish, which were different from those of today's teleosts, evolved into the paired limbs of the tetrapods

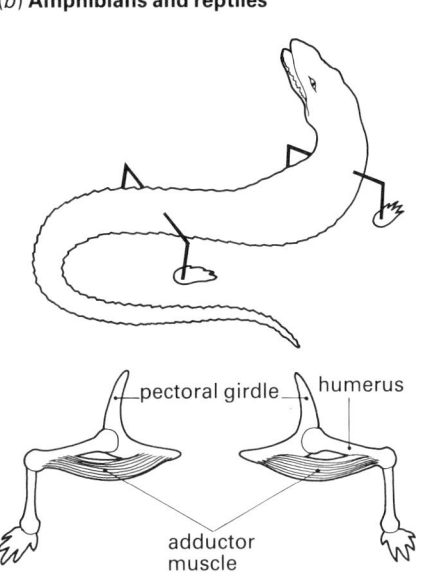

(*b*) **Amphibians and reptiles**

The limbs of these animals are bent and stick out sideways. The body is not held far off the ground, but even this requires well-developed adductor muscles and girdles must be correspondingly large for muscle attachment

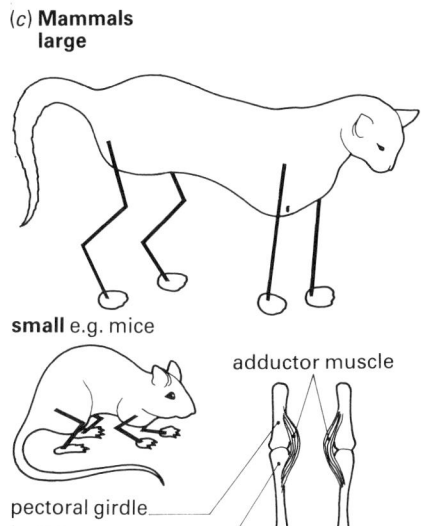

(*c*) **Mammals large**

small e.g. mice

The limb is now rotated and brought under the body, turning through 90° from the reptilian position. The knee is brought forwards and the elbow backwards. The limbs are much straighter, especially in larger mammals. The adductor muscles and the limb girdles are reduced because the weight of the body can be transmitted along what are essentially straight struts

Materials

15 straws (needed to repeat the investigation three times), 50 N Newton meter, scissors, adhesive tape

Procedure

(a) Fix four straws together with a very thin strip of adhesive tape (1 mm). The reason for this is that the tape is acting as a strengthening material. Keep the fifth straw on one side.

(b) Find the breaking point of the single straw by pushing straight down on it onto a Newton meter, as shown in figure 49.

49   Testing the strength of model legs

(c) Repeat (b) three times to get an average result for the single straw.

(d) Repeat procedure (b) three times for the bundle of four straws. For both investigations, note your results carefully.

Discussion of results

1 If the straws represent legs, how does the volume of the larger leg compare with that of the smaller leg?

2 The force causing the straw legs to bend represents the maximum body weight which can be supported by the leg. How does the maximum weight which can be supported by the thinner leg compare with the maximum weight which can be supported by the larger leg?

3 Make a general statement about the size (volume) of legs in relation to weight of body they can support.

4 To support a weight twice that supported by the larger legs, describe the dimensions of the straw model you would require.

5 From your statement in 3, what factors would you expect to limit the size of animals?

6 From your studies so far, can you think of any other factors that would have affected your results?

Show this work to your tutor.

3.4.1 How bones resist environmental forces

As animals increase in size, their skeletal system must enlarge correspondingly. Unless carefully controlled, this could eventually result in animals becoming too bulky to be able to move freely.

Living bone owes its shape and strength to a framework of hard material, parts of which are solid (compact bone) and parts of a looser texture (spongy bone). In general, long bones are compact on the outside with spongy areas towards the ends, but the central marrow cavity is large and contains blood vessels and adipose tissue. In other bones external layers of compact bone enclose a central region of spongy bone containing red marrow, the site of formation of red blood cells. It must be remembered that even compact bone is not really solid.

SAQ 48 Justify the statement that compact bone is not really solid.

Bone tissue is designed to withstand tension and compression. It needs to be strong enough to bear the weight of the body when standing and the forces exerted by that moving weight in locomotion. A general requirement of engineering is for strength with lightness and this is also clearly desirable in animal skeletons. Tubes are stiffer than solid beams of equal strength and are obviously lighter.

Models representing bone can be made from a substance called polymerised resin. If these models are subjected to stresses similar to those experienced by bone and then viewed in polarised light, lines representing the axes of strain, known as trajectories, can be seen in them.

Figure 50(a) shows the pattern obtained from a two-dimensional model of the femur. The force equivalent to body weight borne on one leg acted along the dotted line and bent the femur. Longitudinal stresses were set up in the shaft of the femur but no transverse stresses. The line in the middle of the femur marks the area where no stresses act. Dark fringes on either side mark regular increases of stress. (The numbers indicate relative stress levels.) The stress on the outer face of the femur is a tensile stress. The force by body weight is applied eccentrically and figure 50(b) shows the effect of this in another form.

Study figure 50 and answer the following questions.

SAQ 49 Where are the main regions of stress in a long bone?

SAQ 50 Suggest two reasons why hollow bones are an advantage to mammals and compare this idea with scaffolding and bicycle frames.

Studies on the stress in bones show that there is considerable similarity between the lines of stress and the lines of the trabeculae which provide resistance to the stress. The arrangement of trabeculae is not constant from birth. For instance, if a broken bone is reset badly, the arrangement will change. The trabeculae in the spongy bone at each end of the femur reflect the different stresses acting there. In the hip end two sets of trabeculae occur. One arises from the compressed side to meet the loading forces at the head of the femur, the other comes from the side under tension, crossing the first almost at right-angles and acts to prevent lateral distortion of the femur head. The lower end of the femur is subjected to forces passing through the centre of the shaft and the trabeculae are found in a grid-like pattern. These are shown in figure 51.

50 Stresses on the human femur
(a) Model bone viewed in polarised light

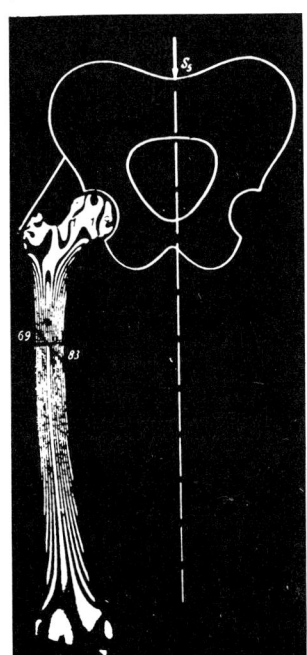

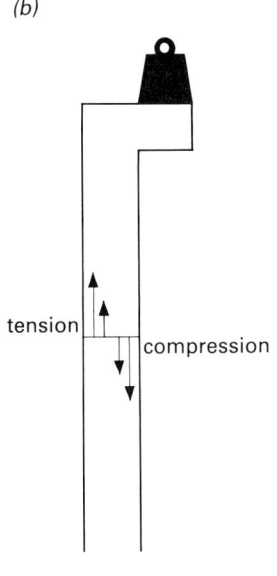

(b)

tension / compression

51 Trabeculae in the ends of the human femur

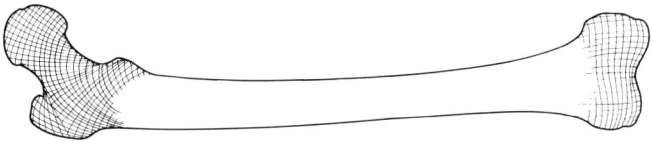

3.4.2 Using braces to reduce stress

Stresses in a structure can sometimes be reduced by making additional forces act upon it. A flagpole is subjected to horizontal forces caused by wind which may bend or even break it. Often guy-ropes are fixed to such poles, the oblique force they apply reducing the bending movement and chances of the pole snapping.

Similarly, the asymmetrical force that acts on the human femur, particularly when weight is on one leg (figure 50) is reduced in life by a brace of muscle and tendon attached to the outer head of the femur. Figure 52 illustrates this and the alteration in stress lines. Compare figure 52 with figure 50.

52 Braces in the human thigh

(a) This model, which includes a brace, shows that the tensile stress in the outer edge of the femur has almost disappeared and the compression stress in the inner edge is much reduced (c.f. figure 50(a)

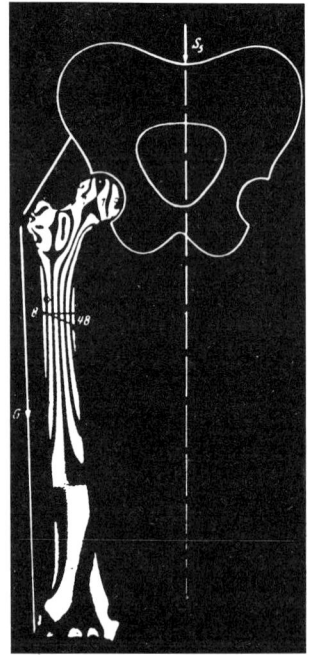

(b)

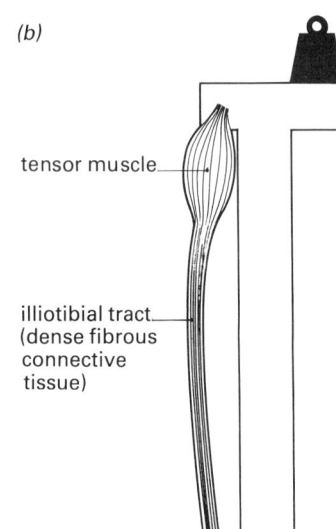

tensor muscle

illiotibial tract (dense fibrous connective tissue)

Examine figure 52.

SAQ 51 What is the chief component of the brace system for the femur?

SAQ 52 Why does its composition particularly suit it for this purpose?

SAQ 53 What added advantage might the tensor muscle give this brace system?

3.4.3 Counterbalancing

Another way in which the stress caused by asymmetrically distributed weight can be overcome is by counterbalancing. The model in figure 53(a) is subject to stress due to the asymmetrically placed weight **A**. If a second weight **B** is placed as a counterweight, as in figure 53(b), stress is greatly reduced. In most mammals, the weight of the body is

53 Counterbalancing

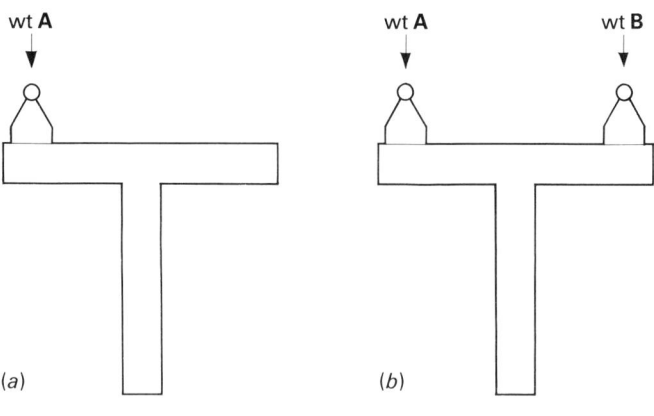

(a) wt **A**

(b) wt **A**   wt **B**

balanced about the main weight-bearing limbs. This may involve structures such as the tail of the kangaroo (figure 54).

54 Counterbalancing in the kangaroo

3.4.4 The role of cantilevers in mammalian support

In mammals, the limbs and muscles carry the whole weight of the body. The hindlimb is jointed to the pelvic girdle in the region of the sacrum. The forelimb is not directly jointed to the vertebral column but weight is transmitted via muscles.

There are similarities between the means of distributing the weight of the body on the legs and the construction of certain bridges, called cantilever bridges. You must be careful not to take the analogy (similarity) too literally since animals are more complex and have additional stress because they move.

A girder supported from a strut develops compression and tension stresses, as shown in figure 55.

55 Compression and tension stresses in a girder supported at one end

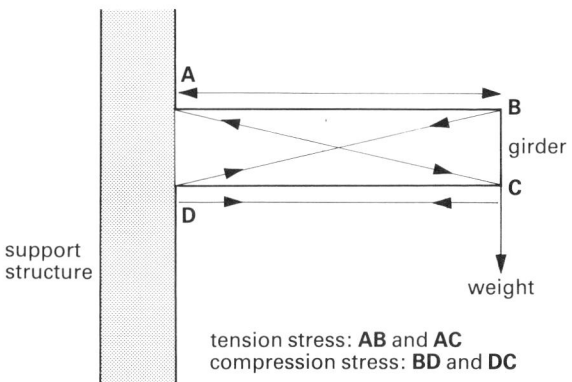

tension stress: **AB** and **AC**
compression stress: **BD** and **DC**

A girder supported at one end is known as a **cantilever**. It is generally constructed of components following the opposing lines of compression and tension stress. This gives it considerable strength.

56 Achieving support with a cantilever girder

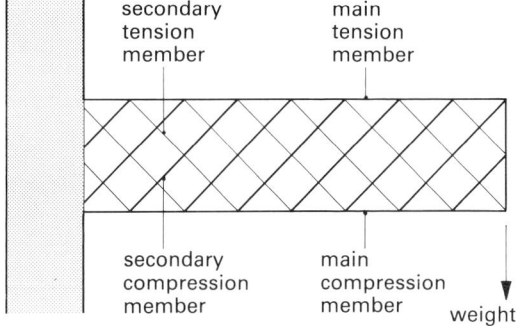

The backbone of a quadruped mammal resembles a system of cantilevers, with the limbs acting as supporting structures and sections of the vertebral column acting as the girders. This is shown in figure 57. The central and neural spines of the vertebrae (see practical G) act as compression members. Muscles and ligaments act as tension members.

SAQ 54 How many cantilever girders are there in the animal in figure 57? Briefly describe each one.

In some mammals, like the one shown in figure 57, the weight is distributed more over one pair of legs than the other. The wolf, on which the illustration is based, supports about 60% of its weight on its forelegs. This accounts for the absence of neural

57 Cantilevers in a quadruped skeletal system

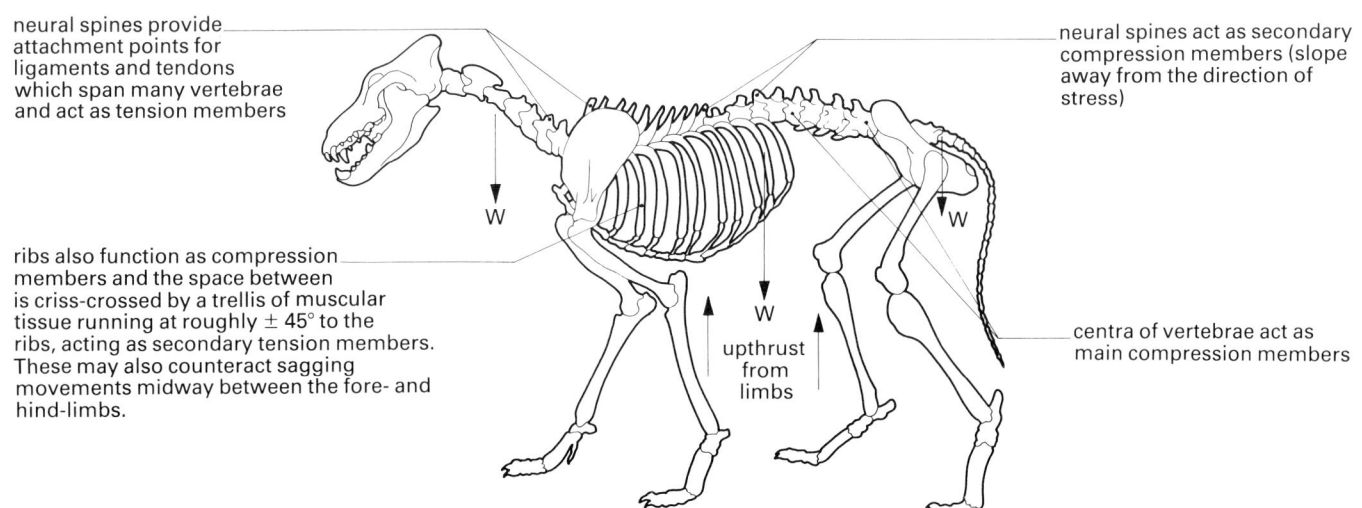

neural spines provide attachment points for ligaments and tendons which span many vertebrae and act as tension members

ribs also function as compression members and the space between is criss-crossed by a trellis of muscular tissue running at roughly ± 45° to the ribs, acting as secondary tension members. These may also counteract sagging movements midway between the fore- and hind-limbs.

neural spines act as secondary compression members (slope away from the direction of stress)

centra of vertebrae act as main compression members

spines on the sacral and caudal vertebrae, and the very prominent projections in the thoracic region.

In a wallaby, support is largely based on a single girder associated with the hindlimbs.

It must be appreciated that the main centres of support may vary as an animal moves.

3.4.5 The importance of arches in support

If you make two model bridges from blocks and a sheet of card, as shown in figure 58, the strength of each model can be gauged by adding weights to the bridge span until it collapses.

58   Model bridges: (*a*) flat bridge: (*b*) arched bridge

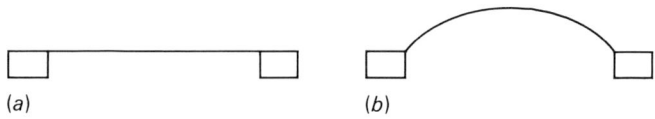

(a)   *(b)*

The results of such investigations will show that the arched model is much stronger than the flat model. If you have never done this, you may like to try it.

Examination of the mammalian skeleton shows that arches are used in it which will provide additional support. Figure 59 shows the hip region of the

59   Hip region of a human skeleton

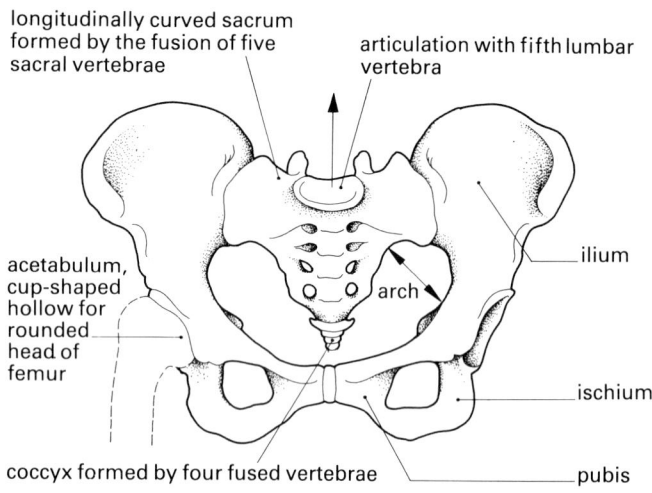

longitudinally curved sacrum formed by the fusion of five sacral vertebrae

articulation with fifth lumbar vertebra

acetabulum, cup-shaped hollow for rounded head of femur

arch

ilium

ischium

coccyx formed by four fused vertebrae

pubis

human skeleton. The sacrum and ilium form an arch which acts to support the weight of the body and transfer it to the legs. The pubis acts as a tie-beam to prevent spreading of the arch under the weight of the body.

Practical G: Examining skeletons

A skeleton, as you know, is made up of many different bones. The bones are held together in the living vertebrate by ligaments, and muscles are attached to the bones by tendons. These allow movement to occur and serve to hold the bones in position.

60   Human skeleton

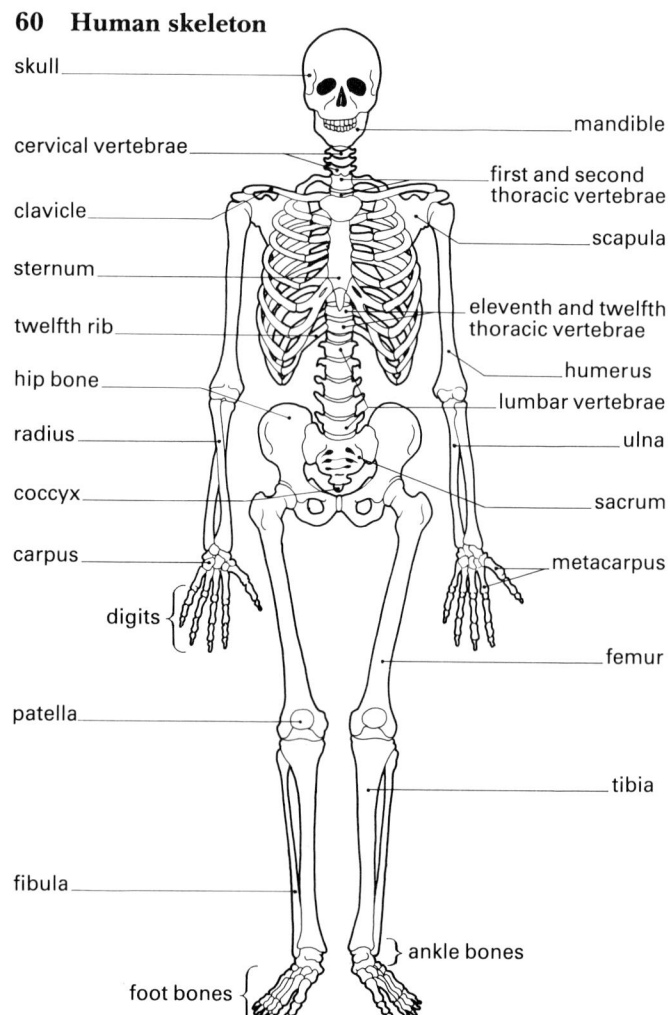

skull

cervical vertebrae

clavicle

sternum

twelfth rib

hip bone

radius

coccyx

carpus

digits

patella

fibula

foot bones

mandible

first and second thoracic vertebrae

scapula

eleventh and twelfth thoracic vertebrae

humerus

lumbar vertebrae

ulna

sacrum

metacarpus

femur

tibia

ankle bones

34

Materials

Mounted skeleton of a quadruped, such as rabbit or rat, articulated skeleton of a human, separate bones of rabbit or rat

Procedure

(a) Examine the mounted skeleton of the quadruped. Use figure 60 to help you identify the different bones.

(b) Identify the main regions of the vertebral column: cervical, thoracic, lumbar, sacral and caudal (tail). Notice how the vertebrae differ in each section and how many there are of each type. Examine separate vertebrae to make this clearer.

(c) Compare the separate vertebrae with figure 61 to identify the main parts of the bones.

(d) Draw and label a thoracic and lumbar vertebra. Annotate to explain their role in support. Notice the direction of the neural spines and explain why it is different in the two vertebrae (cf. secondary compression members of a cantilever girder).

(e) Examine the articulated human skeleton. Use figure 60 to help you to identify the different bones.

(f) Notice the relationship between the vertebral column, the pelvic girdle, the hindlimbs and the feet.

61   A generalised vertebra

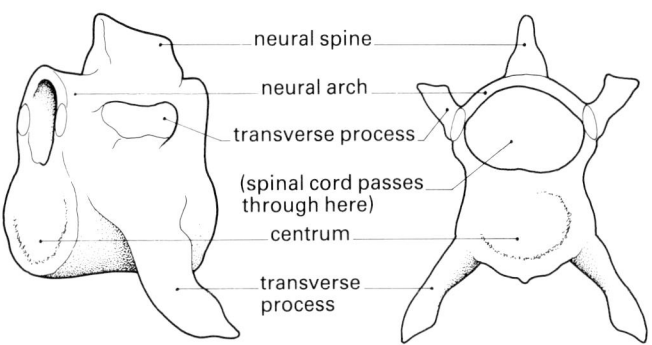

neural spine
neural arch
transverse process
(spinal cord passes through here)
centrum
transverse process

Make a drawing to show this. Label it and annotate it to explain the role of these bones in support.

Show this work to your tutor.

Self test 3, page 111, covers sections 3.2–3.4 of this unit.

3.5 Summary assignment 4

Draw up a table to summarise the role of the factors shown in figure 62 in mammalian support. The first section has been started for you. You will add to this table in the next summary assignment.

62   Factors involved in the support system of a mammal

Factor	Requirement	Role (and example where relevant)
limb bones (long bones)	(a) weight support	compact bone forms long tube – strength with lightness
	(b) stress resistance	arrangement of trabeculae at ends of bone helps resist stresses
	(c) locomotion	length and thickness of bone are related to mass of animal
centre of gravity counterbalancing limb positioning braces cantilevers arches		

3.6 Vertebrate limbs

The bones of all mammalian limbs, indeed of all vertebrate limbs, follow a basic pattern. This is based on three long bones and a collection of smaller bones arranged to form a hand or foot with five digits (fingers or toes). This basic model is described as the **pentadactyl** limb.

The names of the individual bones are different according to whether they are from the forelimb or the hindlimb. The pentadactyl limb is illustrated in figure 63 together with the two sets of bone names.

63 The pentadactyl limb (right) (forelimb names are first, hindlimb names are in brackets)

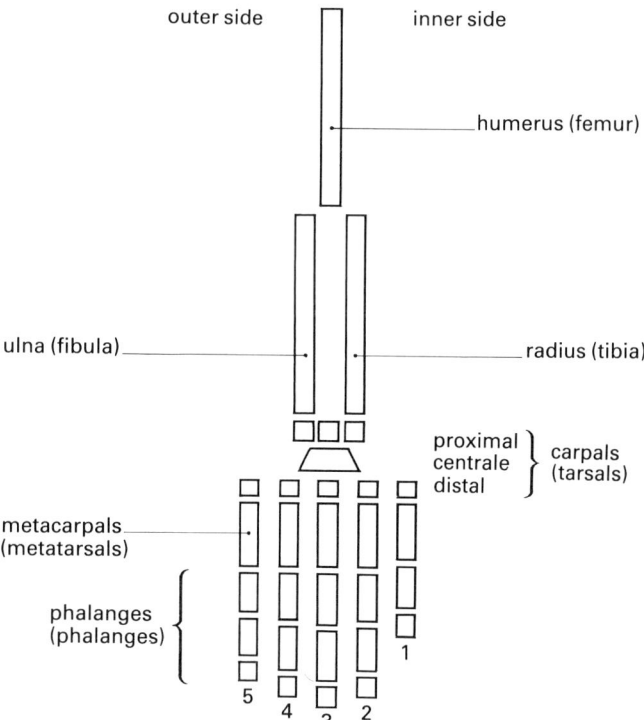

All vertebrate limbs conform to the pentadactyl pattern. Some show changes in the basic pattern. These changes may be small or they may be quite considerable. The wings of birds are modified from the pentadactyl pattern and show many deviations from the basic plan (see figure 121).

Figure 64 shows the bones of a human arm and leg. Study them carefully and then answer the question below.

64 Bones of the arm and leg

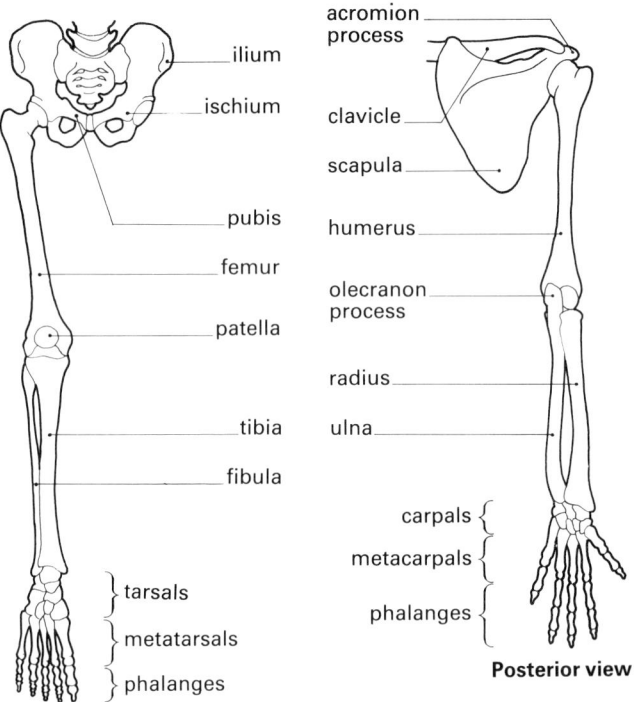

SAQ 55 In what ways, if any, do the human arm and leg differ from the typical pentadactyl plan illustrated in figure 63?

SAQ 56 Study figure 65 and draw up a table to compare the limbs of the mammals. Include relative sizes of component bones, their presence or absence and the function for which the limb is adapted (consider wrist and hand elements under carpals and phalanges only).

65 Variations of the pentadactyl limb (mammalian forelimbs) NB not to scale

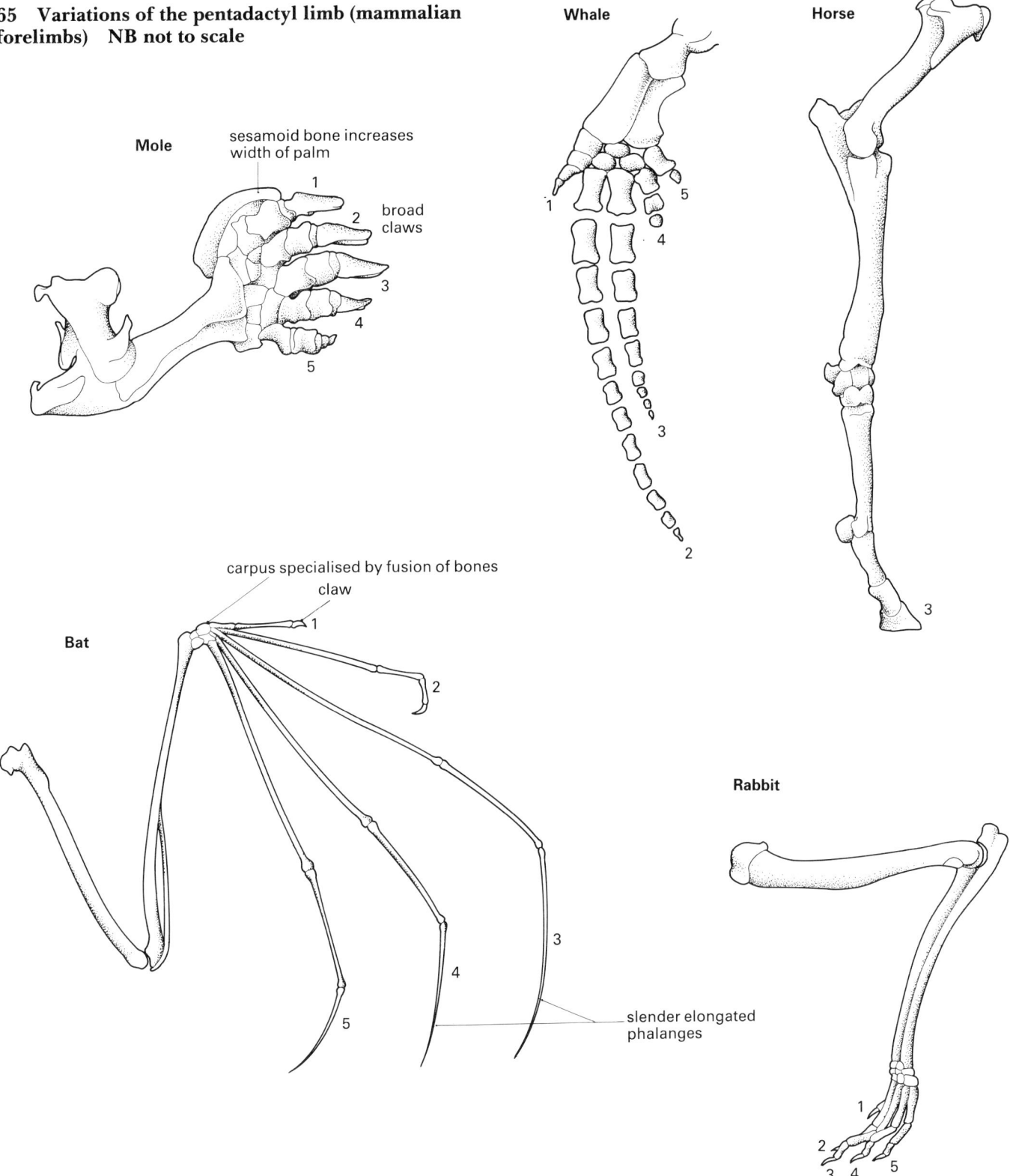

Mole

sesamoid bone increases width of palm

1

broad claws

2

3

4

5

Whale

1

5

4

3

2

Horse

3

Bat

carpus specialised by fusion of bones

claw

1

2

3

4

5

slender elongated phalanges

Rabbit

1

2

3 4 5

66 Components of a mammalian limb (much simplified)

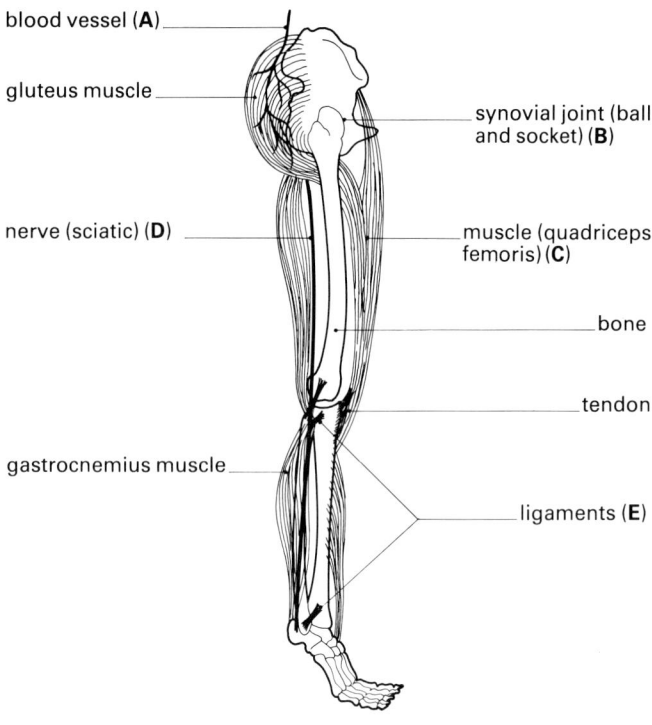

blood vessel (**A**)

gluteus muscle

nerve (sciatic) (**D**)

gastrocnemius muscle

synovial joint (ball and socket) (**B**)

muscle (quadriceps femoris) (**C**)

bone

tendon

ligaments (**E**)

Figure 66 details in a very simplified form the other living tissues that contribute to the functioning of any vertebrate limb.

SAQ 57 Write a brief statement of the role in movement of the structures labelled A–E in figure 66.

3.6.1 Artificial limbs

Accidents or disease can result in the amputation of limbs, but today this need not mean the end of an active life or obvious disfigurement. Modern artificial limbs are lightweight and can give a complete range of normal limb function.

Figure 67(a) shows a man who has had his right leg amputated above the knee. His artificial leg, usually referred to as an above-knee prosthesis, weighs around 2 kg. The shaft which replaced the tibia and fibula consists of a thin-walled stainless steel tube of

25 mm outside diameter. In the future, carbon fibre reinforced plastic may replace stainless steel as it is lightweight and very strong. This strength is essential, because the force on a limb just after heel contact in normal walking is about 1½ times the body weight and this can rise to five times the body weight during running.

Modern artificial legs have been developed after detailed analysis of the walking process and research into different materials.

Movement of the artificial leg occurs because it behaves like a complex pendulum with momentum taking the limb forwards and upwards. Damping devices are needed to reduce excess movement. These mechanisms may be hydraulic, pneumatic or work through mechanical friction. Lubrication is supplied by aircraft bearings made from molybdenum disulphide. Though the knee joint shown in figure 67 functions well, it is not as efficient as a synovial joint. Various knee controls are available to suit the needs of the user and include knee-locking devices and those which assist extension.

Ankle joints are made mainly of nylon and are able to produce all natural ankle movements. Even flexion of the toes can be brought about by rubber springs which absorb energy and release it when needed.

Controlling such a limb requires practice and learning. Older and less active patients may be fitted with simpler models to meet their more limited requirements, with less freedom of movement and even lighter weight.

The mechanical part of the limb is covered with lightweight polyurethene foam, contoured to match the healthy limb. This is covered with a removable flexible silicon 'skin' in a range of colours. It is durable and washable and resistant to stains and grease, and moistured to give as natural an appearance as possible and thus give confidence to the wearer. The finished effect can be seen in figure 67(b).

67 An artificial limb

(a)

(b)

3.7 Joints

A joint is a place where two bones meet. Joints found in the mammalian body are not all of the same kind. There are three main types which differ both in their structure and their function. They are known as sutural joints, cartilaginous joints and synovial joints. In a **sutural joint**, the two bones are held together by fibrous connective tissue. Figure 68 shows sutural joints in the skull where they are most common. These joints look as though they have been sewn together. The word suture comes from the Latin word meaning to sew.

68   Sutural joints in the skull

During ageing, the fibrous tissue in the joint is gradually replaced by bone. Little or no movement is possible at sutural joints.

SAQ 58 What types of fibre would you expect to predominate in the sutures? Give reasons for your answers.

Some joints, such as those between the individual bones of the vertebral column and that between the two parts of the pelvic girdle at the pubic symphysis, contain a disc of cartilage. These are known as **cartilaginous joints**. The cartilage acts as a cushion between the end of the two bones of the joint. The bones are held together by connective tissue ligaments. Limited movement is possible at these joints. A cartilaginous joint is illustrated in figure 69.

SAQ 59 What characteristic features of cartilage make it suitable for these joints?

Joints which allow fairly free movement between bones are called **synovial joints**. In these joints, the ends of the bones are held together by a capsule of ligaments which encloses a small joint cavity. These ligaments arise from and merge with the periosteum of the bones. They hold the bones together firmly but are flexible enough to allow movement to occur. The outer part of the ligament capsule is densely fibrous. The inner part, called the **synovial membrane**, is more cellular and secretes a synovial fluid. Any structure which has parts moving against each other is liable to wear and tear due to friction. In

it is deposited in the joints as sodium mono-urate. These deposits cause intense pain. The condition is commonly known as gout. The synovial fluid also acts to prevent overheating due to friction which might occur at joints during periods of great activity.

In synovial joints, the ends of the bones, known as the **articular surfaces**, are covered by smooth, hard bone which is, in turn, covered by smooth hyaline cartilage. The cartilage helps spread the force on the joints, so reducing the stress.

The features of a synovial joint are summarised in figure 70.

69   A cartilaginous joint

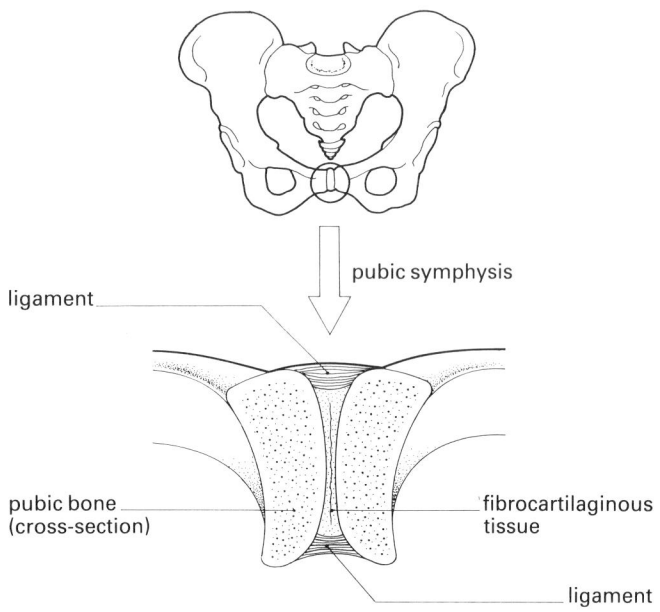

70   A synovial joint

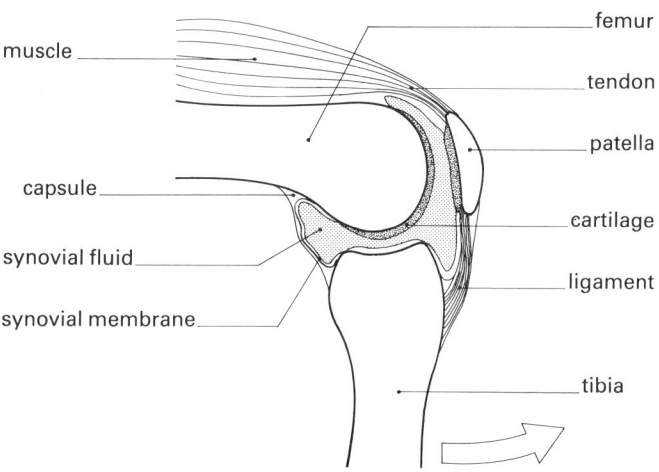

There are different kinds of synovial joint depending on the type of movement allowed. These are illustrated in figure 71.

engineering, this problem may be overcome by greasing the joints. A similar solution is found in joints of living organisms. The synovial fluid is an excellent lubricant. It contains the polysaccharide hyaluronic acid and the protein mucin.

In an experiment, enzymes were injected into the ankles of rabbits to break down the hyaluronic acid. After only 30–48 h of subsequent exercise, the rabbit's joints showed considerable evidence of wear. If excess uric acid is present in the blood of humans

71   Types of synovial joint

(*a*) **Gliding joints** in which the two articular surfaces are flat and slide across each other

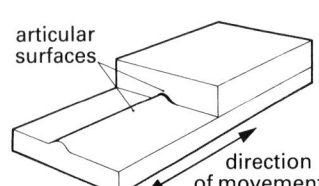

(*b*) **Hinge joint** in which one articular surface is concave and the other is convex, movement is limited to one plane

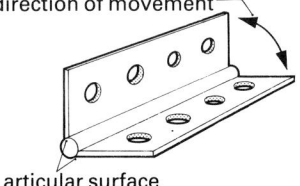

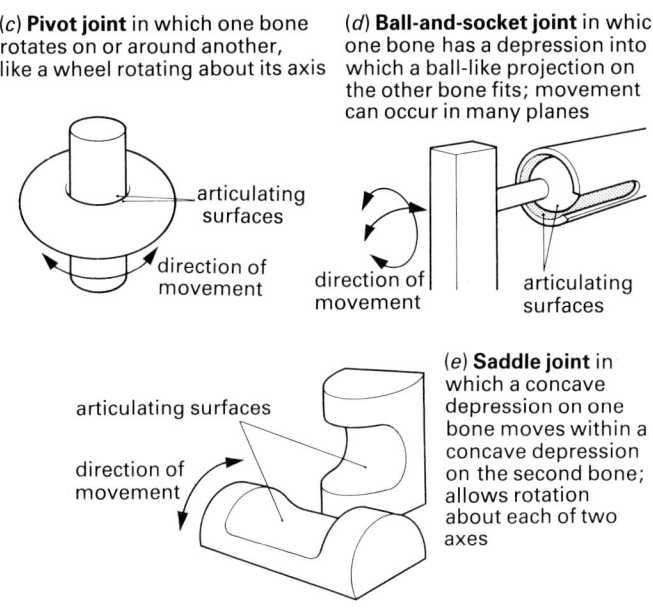

(c) **Pivot joint** in which one bone rotates on or around another, like a wheel rotating about its axis

articulating surfaces

direction of movement

(d) **Ball-and-socket joint** in which one bone has a depression into which a ball-like projection on the other bone fits; movement can occur in many planes

direction of movement

articulating surfaces

articulating surfaces

direction of movement

(e) **Saddle joint** in which a concave depression on one bone moves within a concave depression on the second bone; allows rotation about each of two axes

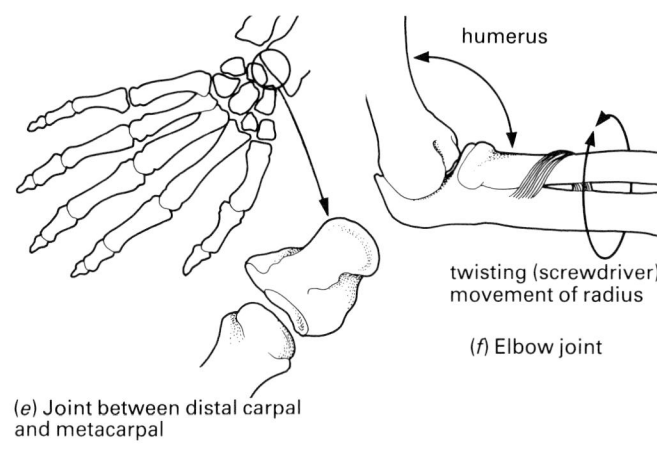

humerus

twisting (screwdriver) movement of radius

(f) Elbow joint

(e) Joint between distal carpal and metacarpal

SAQ 60 For each of the drawings (a)–(f) in figure 72, state which type of synovial joint is represented.

Figure 72 shows drawings of six joints from the human skeleton. Study each carefully. Familiarise yourself with the movements involved at each joint by testing the joints of your own body.

72 Examples of synovial joints from the human body

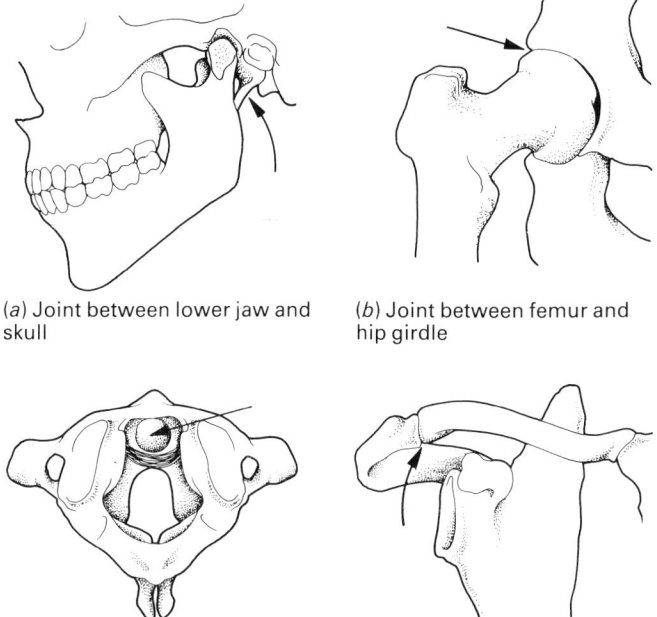

(a) Joint between lower jaw and skull

(b) Joint between femur and hip girdle

(c) Joint between first and second vertebrae (atlas and axis) joint

(d) Joint between shoulder blade and collar bone

3.8 The importance of muscles and the nervous system in support

You have seen that the leg is made up of three bones, the femur and the tibia–fibula, which are jointed at the knee. The great muscle at the front of the leg, the **quadriceps femoris**, is involved in keeping the leg straight. This muscle and the patellar tendon (see figure 73) are richly supplied with sensory nerve endings called **proprioceptors**. If the knee bends, the muscle is stretched and the tendon pulled and this stimulates the proprioceptors. Nerve impulses are set up which pass to the central nervous system and back to the muscle to cause it to contract and thus prevent collapsing at the knee.

Proprioceptors, which are distributed throughout skeletal muscle and tendons, do not 'adapt' and cease to respond to a constant level of input, unlike most other receptors. The coordinated action of a variety of muscles requires that information constantly passes to the brain about the state of each muscle. If you have had the experience of one or both legs 'going to sleep', you have some idea of the problems of walking without proprioceptors.

73   The muscles involved in flexing and extending the leg

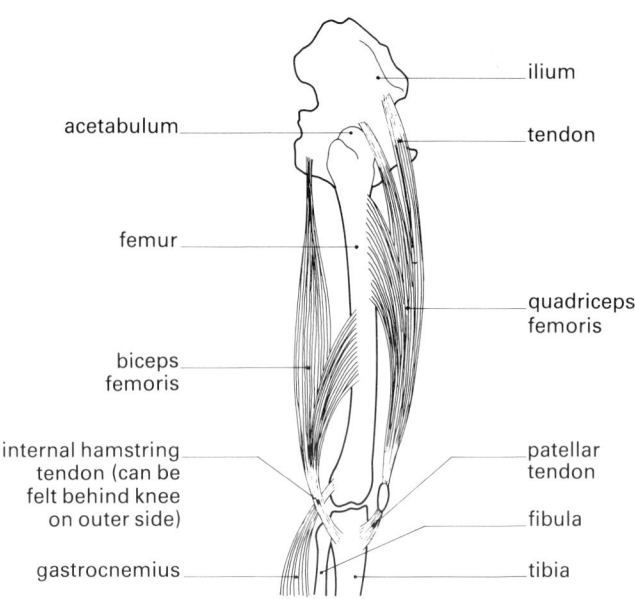

The quadriceps femoris is just one of a group of **antigravity muscles** which help keep the body upright. Further information on the reflex action which causes the knee to jerk and the leg extend can be found in unit 6 *Response to the environment*.

Skeletal muscle tissues associated with voluntary movement are usually elongated cigar-shaped structures. The shape of the muscle is created because each fibre which makes up the muscle is attached to the tendon, and the tapering effect at each end is caused by some fibres joining the tendon nearer to its end than others. The point at which a muscle is attached to a bone by a tendon at its **proximal** (nearest to body centre) end is called the **origin**, the **distal** (furthest from the body centre) attachment is termed the **insertion**. When a muscle contracts, either or both ends of the muscle may pull on bones to initiate movement.

Flexion (or contraction) is the only active movement a muscle can make. It cannot actively stretch again. Therefore, movements which involve opposing action, such as the bending and straightening of the leg, must be carried out by different muscles. Skeletal muscles are usually arranged in pairs, each muscle performing the opposite (antagonistic) action to its partner. Thus, the contraction of one muscle of an **antagonistic** pair will bring about the relaxation (stretching) of the other muscle. (A degree of antagonism can also come from elastic tissue returning to its original length after stretching, but this is less important in vertebrate skeletal muscle systems). Look again at figure 73.

SAQ 61(*a*) Name the muscle whose contraction will cause the leg to bend.
(*b*) Explain how this antagonistic pair of muscles functions to allow the knee to extend and flex.

3.8.1 Coordination and control of skeletal muscle

Section 2.8.1 of this unit explained the relationship between nerve and muscle fibres and the mechanism by which individual muscle fibres are stimulated to contract. The contraction of skeletal muscle is controlled by the nervous system. If impulses fail to reach a muscle it becomes paralysed. Cardiac muscle contracts spontaneously, it is **myogenic**, and smooth muscle may also contract without stimulation. Both these types of muscle receive innervation from the sympathetic and parasympathetic systems (see unit 6, *Response to the environment*) but their effect is to modify the rate and/or strength of the contractions.

This section studies the characteristic functioning of **structural units** of skeletal muscle, that is, of a complete muscle made up of many **motor units** each supplied by a single nerve fibre and capable of functioning independently. The greater the number of motor units that are functioning, the stronger will be the muscle contraction. Where very precise control is required, for example in external eye muscles and the muscles of the larynx, the motor unit may be between 3–15 fibres per motor neuron. This contrasts with human calf muscles where a single motor unit may control over 2000 muscle fibres.

Muscle fibres and their contraction show many of the characteristics of nerve fibres when transmitting impulses.

SAQ 62 Explain what is meant by each of the following when applied to the transmission of an impulse along a nerve fibre.
(a) Resting potential (b) Action potential
(c) Latent period (d) All-or-none event
(e) Depolarisation (f) Summation
(g) Threshold (h) Refractory period

All these phenomena occur in muscle fibres also. In the muscle fibre the electric energy of the action potential is converted into the mechanical energy of contraction. Biologists have come to understand much of the physiology of muscle contraction by using a nerve–muscle preparation. The nerve is stimulated electrically and the subsequent contractions are recorded on a kymograph – a revolving drum covered with recording paper. The apparatus used in such an investigation is shown in figure 74.

The preparation most commonly used is the

74 (a) View of the apparatus used for investigating a nerve–muscle preparation; (b) close-up of a frog nerve–muscle preparation

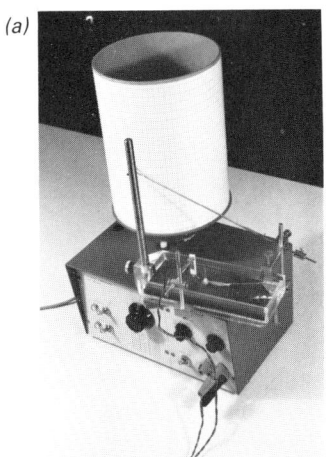
(a)

The gastrocnemius muscle and its innervating sciatic nerve are removed from a pithed frog. The nerve–muscle preparation is placed on a board and kept moist with frog Ringer's solution. The muscle is fixed at one end and the other is connected via a thread to a lever writing on a kymograph drum

(b)

gastrocnemius (calf) muscle of a frog together with the sciatic nerve which supplies it.

SAQ 63 Suggest a reason why an amphibian is often the source of the nerve–muscle preparation rather than a mammal. (Consider the metabolism of these groups.)

75 Kymograph trace from a single stimulus

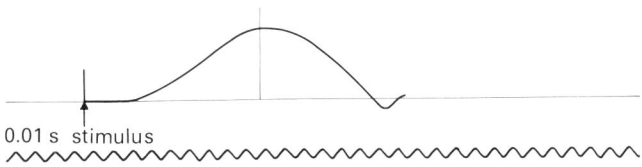
0.01 s stimulus

Figure 75 is a record of a simple **muscle twitch**, the results of a single stimulus above threshold strength.

SAQ 64 Copy the trace and indicate the following events by labelling
(a) the latent period;
(b) phase of contraction;
(c) phase of relaxation;
(d) mechanical bounce.

SAQ 65 How long did the response last?

SAQ 66 Give an explanation of the gradual build-up of response in the period of contraction.

SAQ 67 Write a single sentence describing a simple muscle twitch.

76 The effect of applying two stimuli at decreasing time intervals: (a) 0.02 s; (b) 0.1 s; (c) 0.3 s

S indicates stimulus

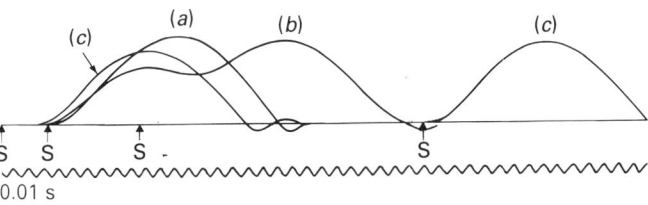
(c) (a) (b) (c)
S S S - S
0.01 s

Figure 76 shows the effects of applying two similar stimuli to the nerve–muscle preparation at different time intervals: (a) 0.02 s interval, (b) 0.1 s interval and (c) 0.3 s interval.

SAQ 68 Describe each response and attempt to explain the response. Which response appears to show summation?

77   Effect of increasing the frequency of repeated stimuli: (*a*) 1 per s; (*b*) 6 per s; (*c*) 10 per s; (*d*) 15 per s

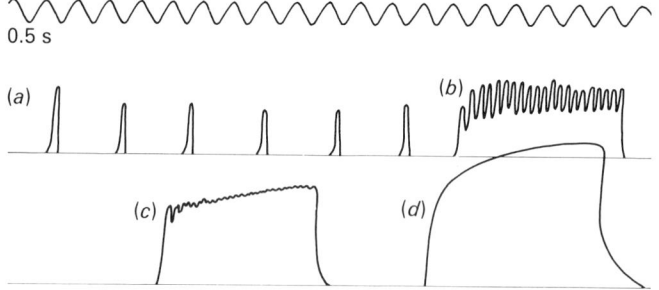

SAQ 69 Look at figure 77. Describe what happens to the responses as the frequency of stimulation increases.

The state of maintained contraction illustrated in figure 77(*d*) is known as **tetanus**. Animal movements under natural conditions, rather than those simulated in laboratory investigations, are not the result of single twitches but a tetanic response. For example, in bending your leg, the flexor muscle (biceps femoris) undergoes a tetanic contraction due to the stream of high-frequency impulses reaching it from the sciatic nerve. The length of time occupied by the flexion depends on the length of time the muscle is stimulated.

A muscle cannot remain in tetanus indefinitely. If stimulation is continued, after a while the response will decline and eventually disappear. This is known as **fatigue**. Fatigue may be caused by several factors, including depletion of supplies of acetylcholine at the motor end-plate and accumulation of lactic acid. The tetanus shown in figure 77 lasted for 3 s, but the period of relaxation due to fatigue which followed it lasted for more than 12 s. Clearly, the time scale would be rather different in a living organism.

Even at rest, most skeletal muscles are in a state of partial contraction, called **tonus**; it is this that maintains posture. Whole muscle relaxation, due to fatigue, is avoided by having only a few motor units activated at any one time. As one set of motor units relaxes, another set takes over.

Much of the control exerted by the central nervous system in support takes place at the level of reflex actions. Then, more complicated coordinated control is added by higher levels of the CNS. Learning is involved, but with frequent use the skills become automatic. The following parts of the brain play an important role in support and movement.

The **cerebellum** receives input from skin, muscles and tendons, from the eyes and from the inner ear where the organ of balance is situated. This part of the brain is essential for controlling the precision of movements. When damaged by injury or disease, movements become jerky and irregular.

The **cerebral hemispheres** are concerned with whole sequences of movement. For example, the spinal cord can organise the movements of running; the cerebellum and related parts of the brain can arrange these movements so that the animal keeps its balance when running; the totality of running is organised by the higher centres in the cerebral hemispheres and learning occurs here.

3.9 Tendons

Tendons which attach muscles to bones may sometimes be quite long. This allows a muscle to act at a considerable distance from the bone that it moves. The **tendo calcaneous** (Achilles tendon) is the common tendon of the gastrocnemius and soleus muscles of humans, and is the thickest and strongest tendon in the body. It is roughly 15 cm long. You can feel it between your calf and heel.

Figure 78 shows that the muscles which move the fingers are at some distance away from them. Move your fingers vigorously and notice the ripples of muscles in the forearm. This positioning of the muscles explains why your forearms ache after using the fingers extensively, such as in typing or playing the piano.

SAQ 70 What is the advantage of the positioning of muscles in figure 78(a) compared to 78(b)?

78   Muscles of the fingers: (a) as they are; (b) as they might look if there were no tendons

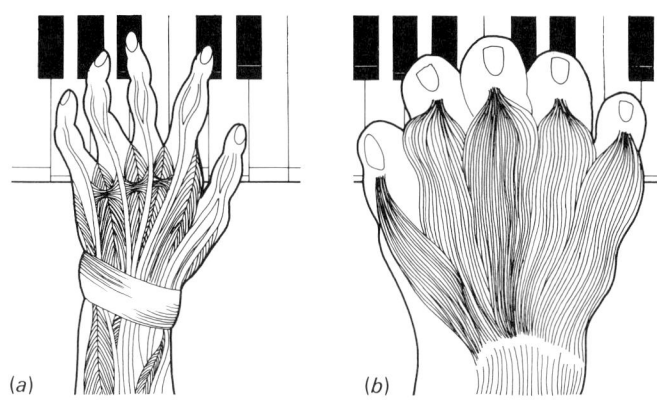

(a)   (b)

3.10 Pre-test: Levers

Answer the following questions. Check your answers with those given on page 132. If you do not get all the answers right, or if you feel unsure of this work, do practical H. If you are happy with your answers, move straight to section 3.10.1.

1 Define the following terms:
lever, fulcrum, effort, load, mechanical advantage.

2 Figure 79 shows different forms of levers, which are labelled **A**, **B** and **C**. Indicate in each diagram which letter (**A**, **B**, **C**) represents the fulcrum, the effort and the load.

3 Describe the three classes of lever. You may use diagrams.

4 Describe the function of the three classes of lever.

Practical H: Levers and their functions

A lever is a rigid bar which may be turned freely about a fixed point. Bones act as levers to bring about movement in the body. There are three types of lever. In this practical, you will examine each type and discover its function.

79   Types of lever

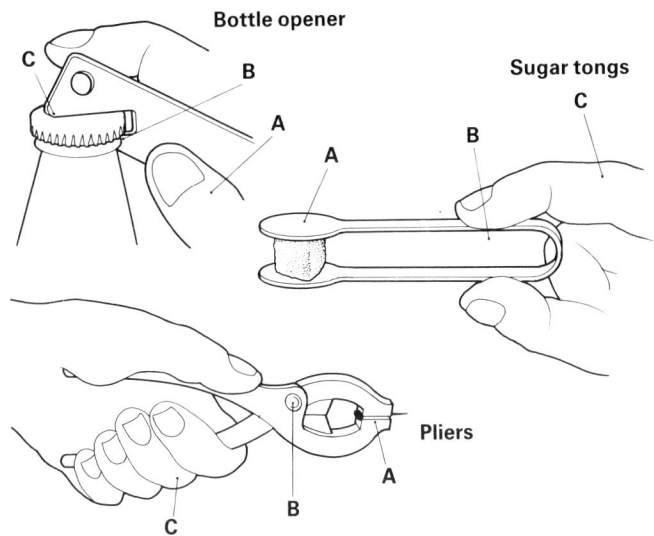

Materials

Metre rule, fulcrum, 500 g mass, Newton meter

Procedure

(a) Set up the apparatus as in figure 80. This represents the first type of lever (first class).

(b) Find the weight of the 500 g mass in newtons.

80   First class lever

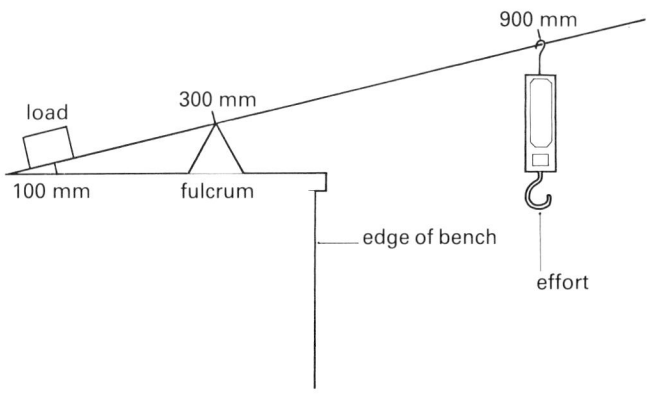

(*c*) Record the effort required to move the load so that the ruler is horizontal.

(*d*) The mechanical advantage (*M*) of a machine is a measure of the relationship between how much effort (*E*) is required and the mass of the load (*R*):

$$M = \frac{R}{E}$$

Calculate the mechanical advantage for this load.

(*e*) Measure (i) the distance which the load moves and (ii) the distance which the effort moves.

(*f*) Record this information in a table (see figure 83).

(*g*) Now set up the apparatus as in figure 81. This represents the second type of lever (second class).

81   Second class lever

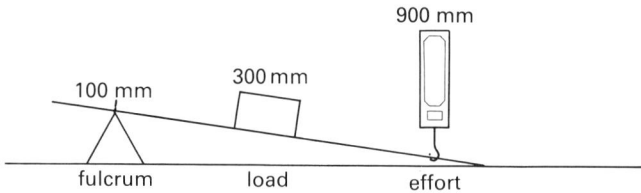

(*h*) Repeat steps (*c*)–(*f*) for this lever.

(*i*) Finally, set up the apparatus as in figure 82.

This represents the third type of lever (third class).

82   Third class lever

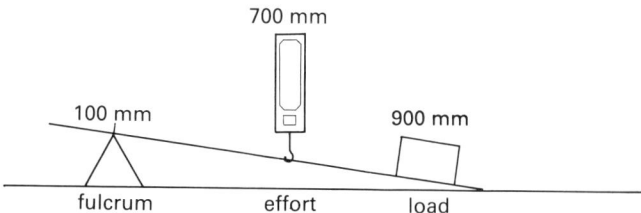

Repeat steps (*c*)–(*f*) for this lever.

Questions for discussion

1 What is the general relationship between the mechanical advantage of a lever and the distances that load moves?

2 What is the function of first and second class levers in terms of their mechanical advantage?

3 What is the function of a third class lever?

Show this work to your tutor.

3.10.1 Laws of levers

The effort required to move a 200 N resistance varies according to: (*a*) how far the effort is from the fulcrum (see figure 84).

84   Relationship between effort and the distance from the fulcrum

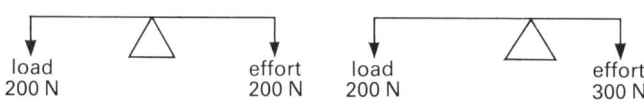

(*b*) how far the load is from the fulcrum (see figure 85).

85   Relationship between load and the distance from the fulcrum

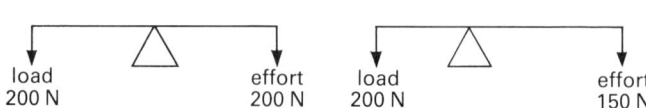

83   Table for results

Lever	Resistance (in N)	Effort (in N)	Mechanical advantage	Distance R moves (in mm)	Distance E moves (in mm)
first class					
second class					
third class					

This relationship is due to the law of levers. This law can be summarised as follows:

$$\text{load} \times \text{distance from fulcrum} = \text{effort} \times \text{distance from fulcrum}$$

SAQ 71 In order to reduce the effort to raise a load, would you move the effort nearer to or further away from the fulcrum?

SAQ 72 In figure 86, if the load was moved from position **B** to position **A**, would the effort at point **C** have to be increased or decreased?

86 Lever for SAQ 72

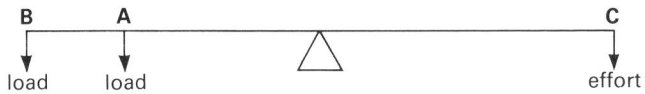

3.10.2 Levers in the body

You are now in a position to relate what you know about levers to movement of the body and its parts. The bones acts as levers. The joints act as fulcra. Muscles supply the effort. The load is represented by gravity acting on the mass of the body or parts of the body. Inertial loads and other non-gravitational loads also occur.

In most lever systems in the body, the distance between the fulcrum and the effort is much less than the distance between the fulcrum and the load. Hence, first and third class levers are the commonest systems in the body. This means that the force applied by the muscles must be very much greater than the load.

Figures 87–9 give examples of three types of lever in the body.

87 First class lever – support of the head

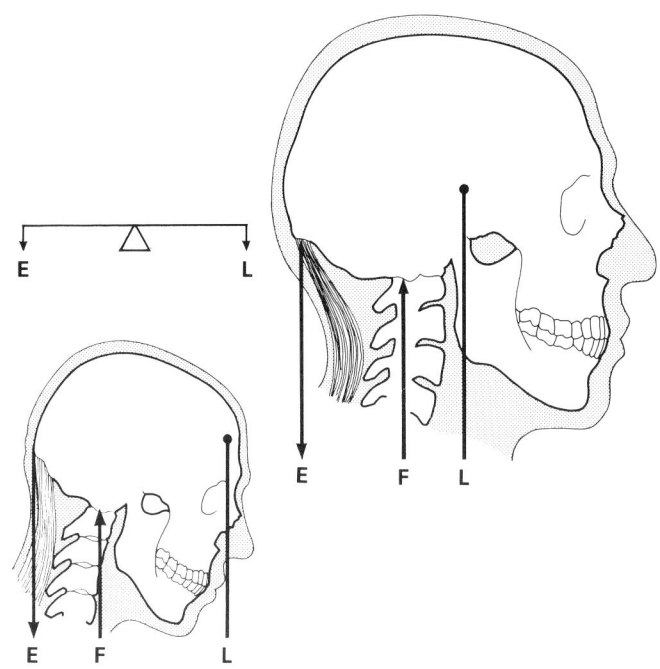

SAQ 73 If you study frequently with your head bent forward, why does the back of your neck feel strained?

SAQ 74 Why do ballet dancers have well-developed calf muscles?

88 Second class lever – standing on tiptoe

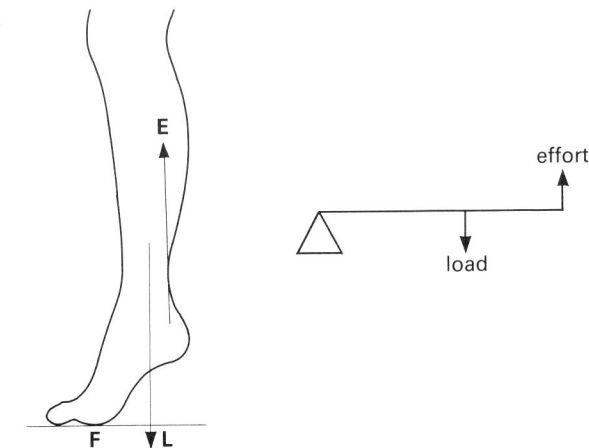

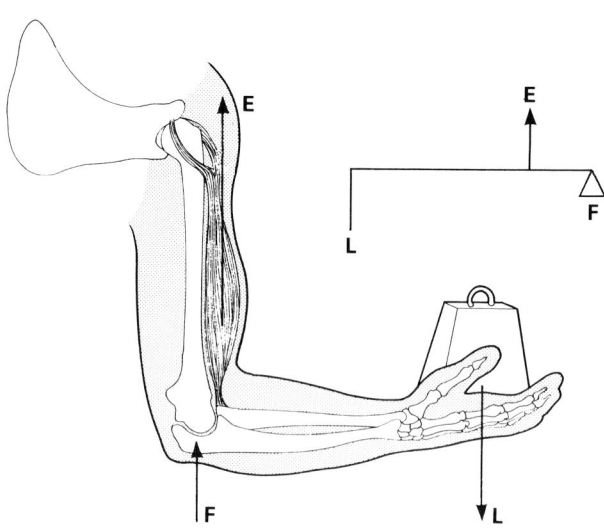

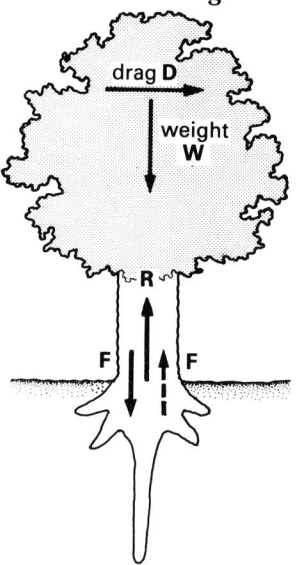

SAQ 75 Why do weight-lifters have bulging biceps?

3.11 Plants and environmental forces

Since higher plants are rooted in the ground, they might not be expected to be subject to as many forces as free-moving animals. Certainly, some plants do grow to greater height and may be of greater mass than any animal, even those supported by water. For example, the largest Californian redwood tree reaches about 90 m in height and its mass is estimated at about 1 million kilograms. In contrast, the blue whale is not more than 30 m long and its mass is around 120 000 kg.

However, plants may be exposed to forces that can uproot them or cause their trunks to break, and they are constructed in such a way that these are rare events. Figure 90 shows the forces that act on trees above ground. Study figure 90.

SAQ 76 What is the cause of the vertical force *W* acting downwards?

The wood of the trunk is compressed by the weight of the tree so that the lower part of the trunk exerts an upward force **R** which helps to counteract the weight of the part above.

SAQ 77 What is the name of the horizontal force?

SAQ 78 When might this force be expected to act?

Experiments on conifers about 8 m high, fixed in wind tunnels, showed that when windspeed was 18 m s^{-1} (approximately 40 mph) drag equalled the weight of the tree (excluding roots). Windspeed in gales would equal 18 m s^{-1}, and some gusts might well exceed this, especially in exposed areas.

Drag forces this strong tend to bend the trunk, stretching the wood on one side and compressing it on the other. These forces set up in the trunk by drag are greater than the drag itself.

The forces marked **F** in figure 90 help to balance the drag.

3.12 Summary assignment 5

1 Complete the table begun in summary assignment 4 by adding information on the following factors: muscles, proprioceptors, nerves, brain, inner ear (refer also to unit 6 *Response to the environment*).

2 Copy the diagram of the pentadactyl limb (figure 63). Annotate the diagram to show how the human arm and leg deviate from the basic plan. For each component of the pentadactyl limb write in the name of the equivalent bone in the fore- and hindlimbs of a mammal.

3 Make sure you have a copy of your answer to SAQ 57.

4 (*a*) Make drawings of a sutural joint, a cartilaginous joint and a synovial joint.
(*b*) Annotate each to explain its structure and function.

5 (*a*) Make sketch diagrams of five named joints in the human body to illustrate gliding, hinge, pivot, ball-and-socket and saddle joints.
(*b*) Annotate each diagram to explain the type of movement involved.

6 Make sketches of figures 75–7. Annotate these to explain their significance.

7 Make brief notes on the use and occurrence of levers in the human body.

8 Make an annotated copy of figure 90 (forces acting on a tree).

Self test 4, page 113, covers sections 3.6–3.11 of this unit.

3.13 Past examination questions

1 (*a*) When a force is applied eccentrically to a bone, stresses of tension and compression are set up. Figure 91 shows a human femur together with regions of tension and compression when an eccentric force is applied to the surface at **A**.
(i) Explain how the gross (*not* microscopic) structure of the entire femur helps it to resist the stresses shown in the diagram.

91 Human femur for question 1

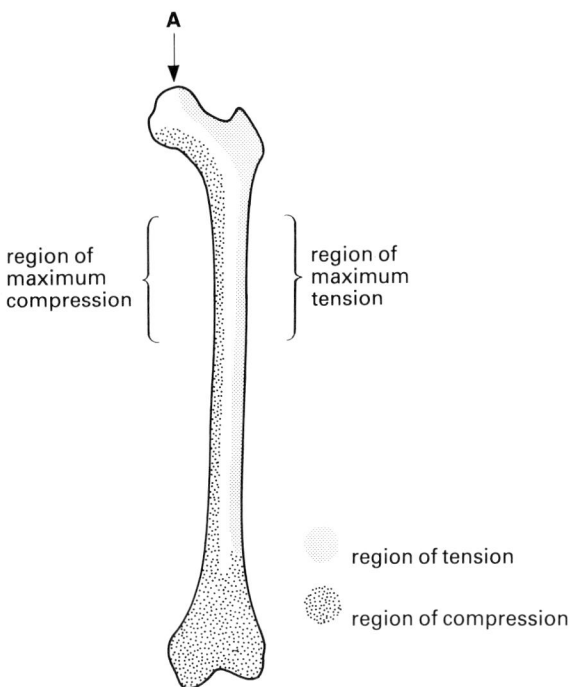

(ii) The stems of herbaceous plants are subject to similar stresses of tension and compression. Explain how *four* structural features of herbaceous stems help them to resist these stresses. (8)
(*b*) Bone tissue must resist tension and compression without breaking and must return to its original shape after stress. Explain how the histological and chemical structure of bone provides these properties. (4)
(*c*) If the concentration of calcium ions in blood plasma falls below a certain level, what is the effect on
 (i) the parathyroid gland?
 (ii) the thyroid gland?
(iii) bone tissue? (4)
(*d*) Apart from its role in bone tissue, calcium has many other functions in mammals and herbaceous plants. Describe any *four* of these other functions. (4)
(AEB, 1982)

2 Figures 92(*a*) and (*b*) are photographs of a life-size model of the skeleton of an iguanodon.
(*a*) Identify the parts of the skeleton labelled **A**, **B**, **C**, **D** and **E**. (5)

(b) Identify the vertebrae labelled **F**, **G** and **H**. (3)

(c) Suggest a function for the two bones marked **X**. (1)

(d) Suggest how the weight of a living iguanodon would have been supported in a balanced position. (2)

(e) Describe how the cerebellum of the brain, the inner ear, and stretch receptors in skeletal muscle interact to keep the human body in an upright position. (9)

(Total 20 marks)

(AEB, 1981)

92 Iguanodon skeleton

(a)

(b)

Section 4   Locomotion

4.1 Introduction and objectives

This section studies locomotion in a selected group of protists and animals. For an organism to move the whole of itself from one place to another it must achieve three things:

1 it must somehow be propelled in the right direction – **propulsion**;
2 its body must be supported so that it can act against the medium in which it is moving – **support**;
3 although there may be a temporary phase of instability or imbalance, it must be stable at the end of the movement – **stability**.

Organisms have achieved these three requisites in a number of different ways and in a manner specially adapted to the medium (water, land or air) in which they move. You already know quite a lot about how support and stability are achieved on land and how musculoskeletal systems are the basis of movement, and these ideas will be developed and extended in this section.

After completing this section you should be able to do the following.

(*a*) Give an account of three types of movement which occur in the Protista.

(*b*) Describe how an earthworm moves.

(*c*) Outline the events of walking in human beings.

(*d*) Indicate briefly some of the differences in tetrapod (quadruped) locomotion.

(*e*) Describe how swimming, stability and buoyancy are achieved in fish.

(*f*) State the problems of movement through air.

(*g*) Describe how gliding and flapping flight are achieved in birds.

(*h*) Briefly describe the structures associated with flight in an insect and explain how they operate.

Extension

(*a*) Outline the range of mechanisms exploited by different animals for movement through water.

4.2 Locomotion of Protista

In this section you will study locomotion in *Amoeba*, *Paramecium* and *Euglena* by means of practically based investigations.

Certain Protozoa, together with some other cells including some vertebrate white blood cells, are able to move by putting out temporary projections of the cytoplasm known as **pseudopodia**. The rest of the cytoplasm then apparently flows into these pseudopodia. Several theories have been put forward to explain how this occurs. Before considering these theories, it is useful to recall some basic facts about *Amoeba* and cytoplasm.

SAQ 79 The ground substance of cytoplasm is a colloid. What is a colloid? In what forms may a colloid exist?

SAQ 80 It is possible to distinguish two forms of cytoplasm in *Amoeba*. Describe them.

An early theory considers *Amoeba* as basically a tube of plasmagel containing fluid plasmasol. The gel is changed to sol by protein molecules in the cytoplasm altering their shape. The folding up of molecules as gel changes to sol at the rear might provide the pressure forcing the fluid sol forward. The elastic properties of the cell membrane would also play some part. Another theory claims that forward movement is not due to pressure at the rear but to a pulling force at the front. There are also theories involving the presence of contractile fibres composed of actin and related proteins. Figure 93 summarises the cytoplasmic changes involved in amoeboid movement, while figure 94 summarises the theories put forward to explain it.

93 Cytoplasmic movements during amoeboid movement

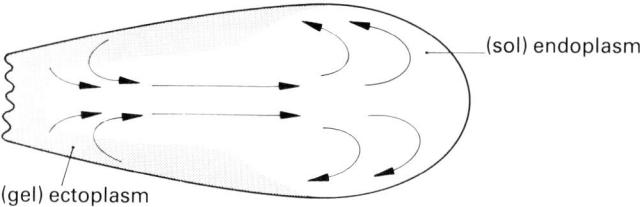

(sol) endoplasm

(gel) ectoplasm

94 Summarising the theories of amoeboid movement

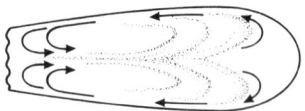

A The posterior region of ectoplasm squeezes the more fluid endoplasm forward through a tube of ectoplasm

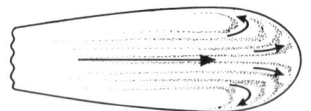

B Cytoplasm at the front end contracts pulling *Amoeba* forwards. The change from endoplasm to ectoplasm is the source of the contraction

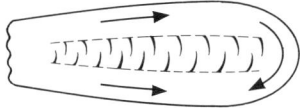

C Recent thinking gives most support to the theory that contraction is a basic property of living cytoplasm depending on the presence of actin-like protein molecules (microfilaments) which slide against each other in a ratchet system related to that seen in muscle contraction

For *Amoeba* to move forward, there must be points where the cytoplasm adheres to the substrate in order to exert a force against it. Figure 95 shows *Amoeba* from the side. New pseudopodia make contact with the substrate and adhere, but the rear and the withdrawing pseudopodia are lifted clear of the substrate.

95 Amoeboid movement (side view)

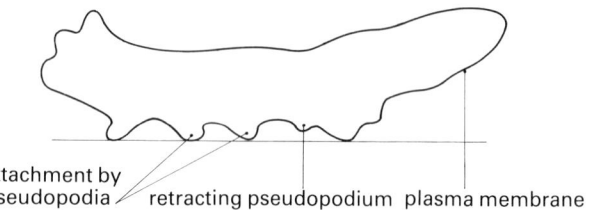

attachment by pseudopodia   retracting pseudopodium   plasma membrane

Practical I gives you the opportunity to observe the phenomenon for yourself.

Practical I: Investigating amoeboid movement

Materials

Culture of *Amoeba*, 2 slides, 2 cover-slips, dropping pipette, monocular microscope and lamp, graph paper or calibrated slide and eyepiece graticule

Procedure

(*a*) Use a pipette to transfer a drop of culture onto a slide.

(*b*) Cover with a cover-slip and observe the movement of an *Amoeba* under low power and high power. Make sketches at intervals and indicate the direction of movement by arrows. Briefly describe what you observe of the cytoplasmic contents.

(*c*) Relate your observations to figures 93 and 94 and your knowledge of the nature of cytoplasm in *Amoeba*.

(*d*) Determine the speed of locomotion of the *Amoeba* by timing how long it takes to move across the field of view of the microscope. Measure the distance of the field of view by the method described in the unit *Inquiry and investigation in biology*, practical C. (Alternatively, use a calibrated slide and eyepiece graticule which your tutor will explain how to use). Express the speed in mm s^{-1}.

(*e*) Make an illustrated report of your findings. Include a consideration of questions 1–4.

Questions for discussion

1 Does the *Amoeba* normally move in a straight line? If not, how does it move?

2 Does it appear to have a permanent anterior and posterior end?

3 What changes could you observe in the cytoplasm of a moving *Amoeba*?

4 Figure 94 summarises the three main theories of how amoeboid movement occurs. What line of investigation could be followed to test one or more of these theories?

Show this work to your tutor.

Cilia and flagella show the same fundamental structure under the electron microscope. They differ in their length, mode of beating and the type of movement this produces. The bending of cilia and flagella is thought to be brought about by a sliding filament mechanism. As well as providing organelles of locomotion for many Protista, they function as important effector structures in most groups of animals and in reproductive cells of some simple plants.

Practical J: Investigating ciliary and flagellar movement

In this practical you will observe movement in *Paramecium* and *Euglena* and relate your findings to what is known of the structure and function of cilia and flagella. The basic structure of these organelles was covered in the unit *Cells and the origin of life*.

Materials

Culture of *Paramecium*, culture of *Euglena*, 4 slides, 4 cover-slips, monocular microscope and lamp, dropping pipette, methyl cellulose solution

Procedure

(*a*) Use a pipette to transfer a drop of *Paramecium* culture onto a slide.

(*b*) Add a little methyl cellulose solution to slow down the activity of the organisms. Cover with a cover-slip.

(*c*) Observe the movement of a *Paramecium*.

(*d*) Relate the movement of the cilia to figures 96 and 97.

96 Cilia structure

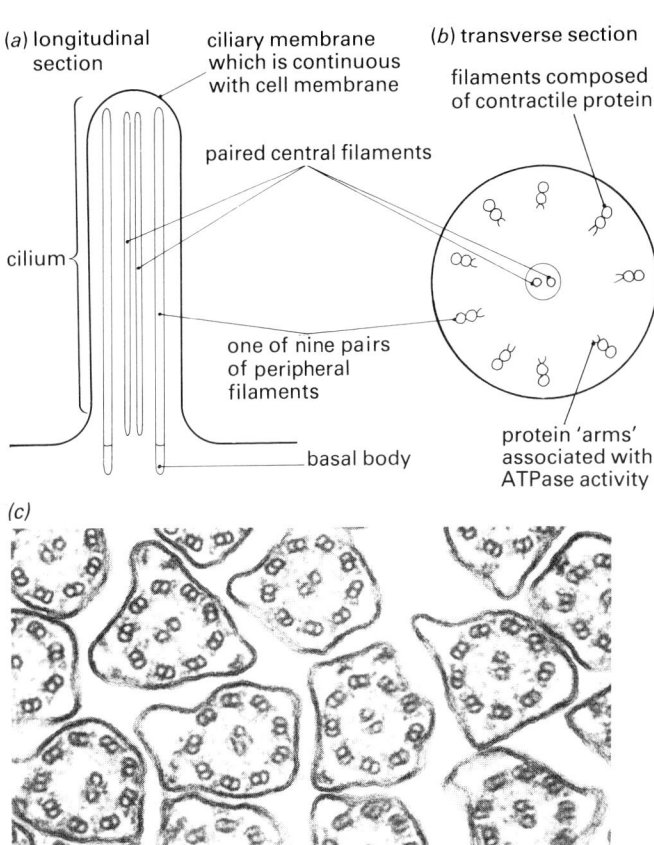

(*a*) longitudinal section — ciliary membrane which is continuous with cell membrane — paired central filaments — cilium — one of nine pairs of peripheral filaments — basal body

(*b*) transverse section — filaments composed of contractile protein — protein 'arms' associated with ATPase activity

(*c*)

97 Cilia action

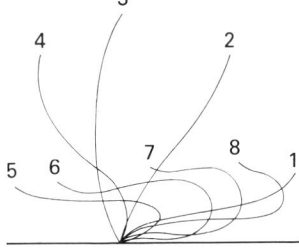

(*a*) Numbers 1–4 indicate the forward effective stroke, the limp recovery stroke is labelled 5–8 and does not bring about significant movement

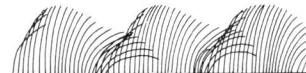

(*b*) Cilia are closely and regularly packed and their beats are coordinated in a pattern called metachronal rhythm which is usually compared to the passage of wind through a corn field.

This reduces turbulence and keeps the fluid medium moving steadily in the opposite direction to which the organism moves

(*e*) Make an illustrated report of your findings. Include a consideration of questions 1–3 and 6.

(*f*) Repeat this procedure for studying movement in *Euglena*. Refer to figures 98 and 99 and to questions 4–6.

98 Flagella and body movements in *Euglena*

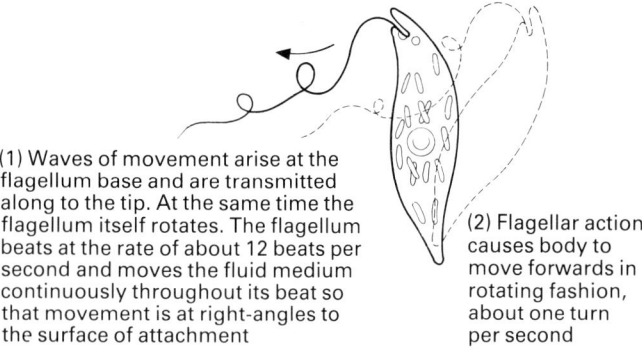

(1) Waves of movement arise at the flagellum base and are transmitted along to the tip. At the same time the flagellum itself rotates. The flagellum beats at the rate of about 12 beats per second and moves the fluid medium continuously throughout its beat so that movement is at right-angles to the surface of attachment

(2) Flagellar action causes body to move forwards in rotating fashion, about one turn per second

99 Sequences of body positions of *Euglena* as it moves from X to Y

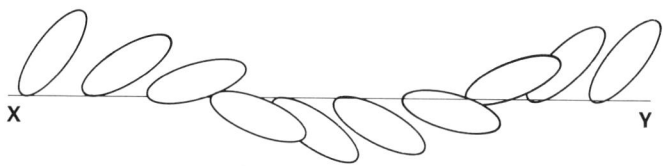

Because the tip of the organism rotates while being pushed to one side, it follows a corkscrew course of movement. This may be compared with the movements of a propeller setting up forces on the water which cause it to be displaced in a forwards direction

Questions for discussion

1 How do the cilia move in *Paramecium*?

2 Do they appear to be coordinated?

3 What do you notice about the direction of movement in *Paramecium*? For example does it swim straight, does it rotate, can it move forwards and backwards?

4 How does the pathway of movement in *Euglena* compare with that of *Paramecium*?

5 Does the body shape of *Euglena* change during movement? If so, what does this indicate about the nature of the cytoplasm?

6 How does the speed of movement compare in the two organisms?

Show this work to your tutor.

4.3 Locomotion of the earthworm

The earthworm is a member of the phylum Annelida. It shows two evolutionary advances over the sea anemone considered in section 1.6.1, but it also depends upon a hydrostatic skeleton. The first advance is that it possesses a **coelom** containing coelomic fluid. This cavity surrounds the alimentary canal, providing the structural basis for the hydrostatic skeleton and allowing for independent movement of the gut. In addition, the earthworm is **metamerically segmented**, that is the body is divided along its longitudinal axis into a series of segments or units, each of which contains elements of some of the chief organs. Figure 100 shows a cross-section through the earthworm. Study this carefully, then proceed to practical K.

100 Section through the body of an earthworm

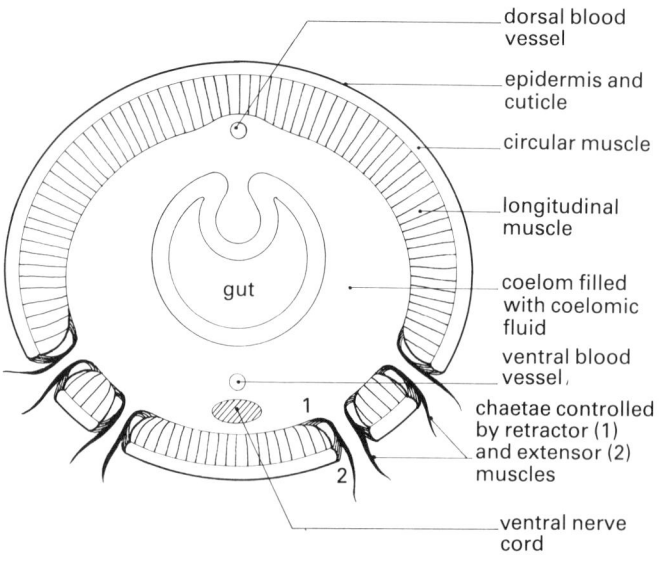

dorsal blood vessel

epidermis and cuticle

circular muscle

longitudinal muscle

coelom filled with coelomic fluid

ventral blood vessel

chaetae controlled by retractor (1) and extensor (2) muscles

ventral nerve cord

gut

Practical K: Investigating locomotion in an earthworm

In this practical you will investigate movement of a living earthworm in relation to its internal structure as seen in a prepared microscope slide.

Materials

Live earthworm in a pie-dish with a little water to keep moist, piece of scrap paper, prepared microscope slide of TS through earthworm, monocular microscope and lamp

Procedure

(*a*) Observe the movements of the earthworm in the dish.

(*b*) Record your findings with annotated diagrams.

(*c*) Allow the worm to move over a piece of scrap paper. Record anything you hear. Alternatively, or in addition, allow the worm to move over your hand. Record what you feel.

(*d*) Observe the slide of the TS through an earthworm under the microscope. Using figure 100, identify the layers of muscle in the body wall, the coelom, the chaetae and their associated muscles.

(*e*) Make an annotated diagram of your slide to show these features. Explain how they are involved in movement.

(*f*) Using findings from your study of the microscope section, annotate your initial diagrams to explain what internal events are occurring which cause the movements you observed.

Questions for discussion

1 What are the roles of the following in earthworm locomotion: body wall, fluid-filled coelom, chaetae.

2 Study figure 101 and use it to explain how the worm moves forward. These drawings are based on a cinematograph record. The track of individual points on the worm's body and their movements relative to each other are shown by the lines running obliquely forwards from left to right of the diagram. It may be helpful to discuss this with others in your group.

3 How is the method of locomotion of the earthworm particularly suited to a *burrowing* animal?

Show this work to your tutor.

4.4 Human locomotion – walking and running

Figure 102 shows four types of movement that occur during the process of walking.

SAQ 81 Study figure 102 and either walk yourself or observe a friend walking. Describe walking by listing the sequence in which the four types of movement occur during walking.

101   Diagram showing the mode of progression of an earthworm

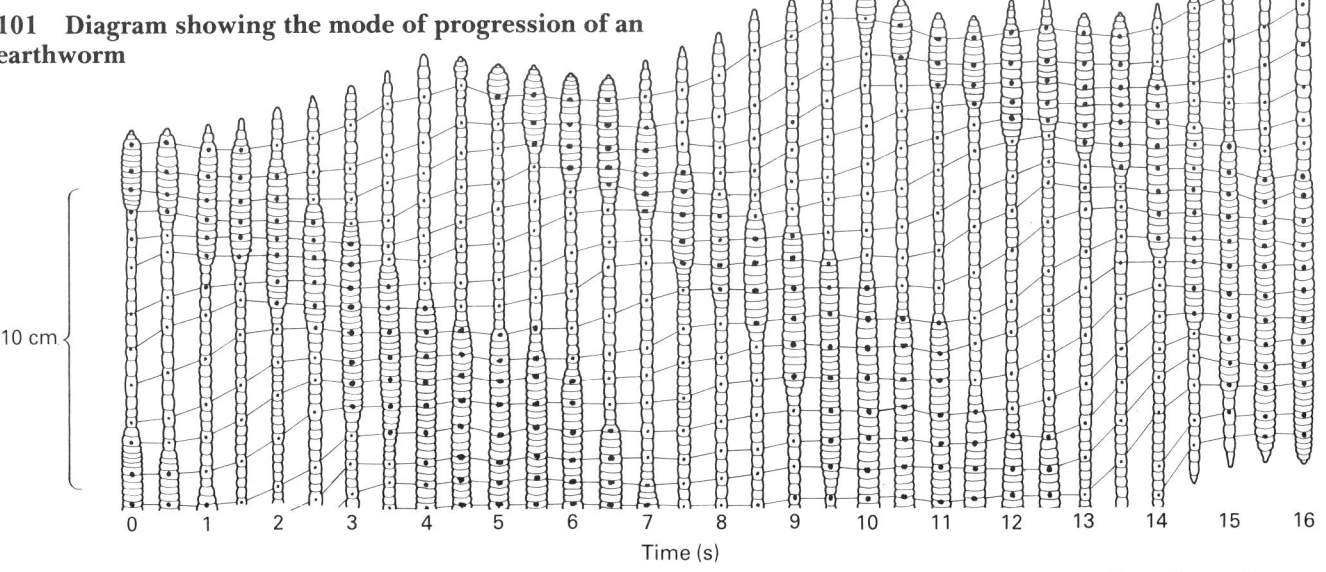

Time (s)

Extending region
circular muscles contracting
longitudinal muscles relaxed
chaetae retracted

Stationary region
circular muscles relaxed
longitudinal muscles contracting
chaetae extended for anchorage and thrust

Retracting region
circular muscles contracting
longitudinal muscles relaxed
chaetae retracted

102   Movements occurring during walking

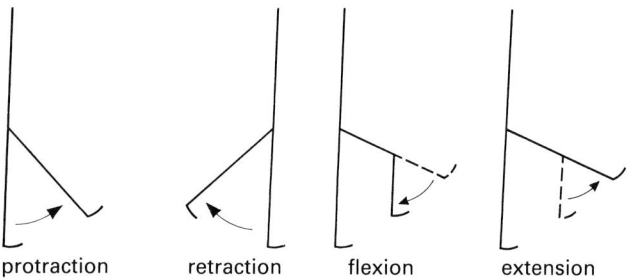

protraction   retraction   flexion   extension

SAQ 82 Which of these stages is directly responsible for propelling the body forward?

Figure 103 shows some of the muscles involved in walking in humans.

103   Muscles involved with human walking

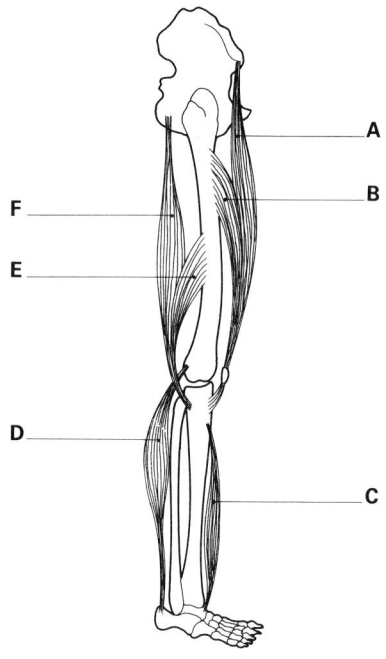

SAQ 83 Using figure 103, state which muscle contracts to bring about each of the stages in figure 102. Remember that some muscles will be forcibly extended at the same time.

During walking, when the foot makes initial contact with the ground, the ankle is flexed, and when the foot leaves the floor, the ankle is extended.

SAQ 84 Which muscles are contracted (*a*) when the ankle is flexed, (*b*) when the ankle is extended? Use figure 103 to help you answer this question.

In addition to the types of movement shown in figure 102, there are three other common types. These are shown in figure 104, together with some of the muscles which cause the movement.

104   Additional types of movement of the leg

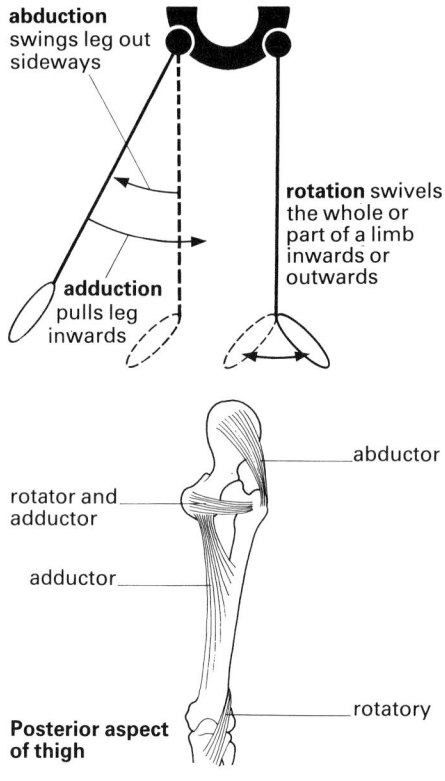

Remember that human beings do not only move their legs when walking. The natural actions of a person's arms and legs are essentially the same as a tetrapod, such as a horse, whose diagonal limbs are working together. The right arm swings forward with the left foot and the left arm with the right foot. This compensates for the inclination of the body to the opposite side. The stronger the push of the leg, the greater the distance swung by the arm.

When walking, each foot is on the ground for more

than half the stride and sometimes both feet are on the ground together. In running, each foot is on the ground for less than half the stride and both may be off the ground at the same time. (Make a point of observing someone running or reflect on your own movement as soon as convenient.)

SAQ 85 If the foot is on the ground for less time during running, will it need to exert a greater or lesser force?

SAQ 86 List the postural differences between a person walking and a person running.

Recent research shows that human walking and running can be viewed as techniques for travelling with the least possible energy cost. In walking, relatively little energy is used by keeping the legs fairly straight so that leg length is fairly constant while the feet are on the ground. In running, the tendons act as springs storing elastic strain energy at one stage of the step which is returned as elastic recoil at another and thus energy is saved. Below a critical speed walking is more economical than running, but above that speed, running is the more economical method of locomotion.

4.4.1 Movement in some quadrupeds

(To begin with, a word about terminology. Both *quadruped* and *tetrapod* are words referring to animals with four feet, and you may find both used in textbooks and examination papers. Quadruped is more often associated with mammals and tetrapod is probably a more general term.)

Figure 105 illustrates the action of the hindlimb when propelling the body forward. The flexor and extensor muscles generate the force for propulsion.

SAQ 87 (*a*) How does the foot act as a lever? Indicate where fulcrum, load and effort will act. (*b*) What are the directions of the forces exerted in figure 105(*b*)?

This picture of movement and forces is very similar to that of human beings though, of course, the human leg is much straighter and has more flexibility at the knee joint.

105   Action of the hindlimb of a rabbit about to leap: (*a*) partially flexed; (*b*) extended

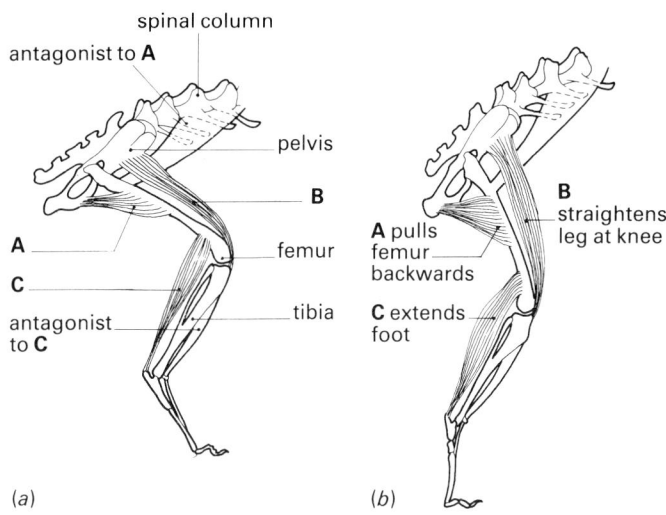

(*a*)   (*b*)

A four-legged mammal may be compared to a table supported at the four corners. If a weight is placed at the centre of the table and one of the legs is removed, then the table will topple over. The position of the weight indicates the centre of gravity. If the weight is nearer one end, one of the legs at the other end may be removed and stability is maintained. If the centre of gravity is towards one end, the animal has two alternative triangles of support, the two legs close to the centre of gravity and one or other of the two legs further away from it.

Among mammals, some have their centre of gravity towards the front like a horse, which can often be seen with one hind leg at ease bearing little weight, while others carry their weight well back. Rabbits, squirrels and bears are in this group and they are often observed rearing up on their hind legs in an almost bipedal position.

During movement, for stability to be maintained, the centre of gravity must shift in order to keep a balanced triangle and only one leg may be off the ground at any time. Observation of tetrapods (even a crawling baby) reveals an identical pattern of movement. Legs are lifted from the ground in a definite order. (Check this for yourself by observing.) This is: right forefoot, left hindfoot, left forefoot, right hindfoot, and so on. This is the only pattern which ensures that no foot is lifted unless the centre of gravity of the body lies over the triangle marked out by the other three. This type of locomotion is rather slow. In more rapid motion, two legs may be off the ground simultaneously but the period of instability is very short. A horse may balance its body alternately on two feet on the same side of the body and on a pair of diagonally opposed feet. This is shown in figure 106.

While trotting, its balance is on diagonally opposed feet (figure 106(b)), and during galloping there are never more than two feet on the ground simultaneously and sometimes the whole of the animal is in the air (figure 106(c)). The power for forward movement in a galloping horse comes almost entirely from limb muscles and relatively little from back muscles. The greyhound shown in figure 107 increases the span of its hindlimb by arching and flattening its vertebral column and the contraction of back muscles greatly supplements the power of the limbs.

Speed in running results from the length of a stride, multiplied by the rate of stride. High speed requires that long strides are made at a rapid rate.

Longer legs take longer strides and animals that run fast have the distal portions of their limbs elongated compared with the parts of the limb closest to the body.

107   Locomotion in the greyhound

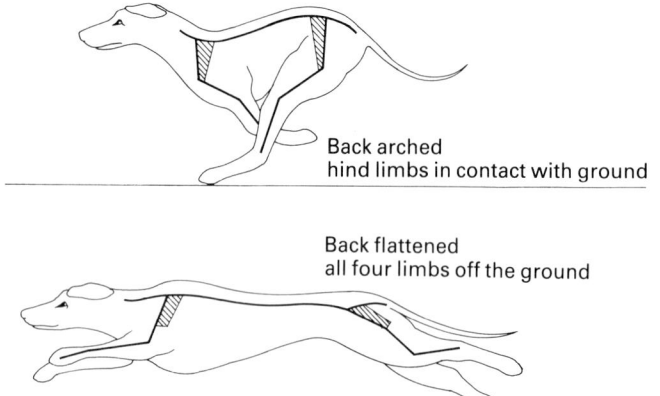

Back arched
hind limbs in contact with ground

Back flattened
all four limbs off the ground

106   Movement of limbs in the horse

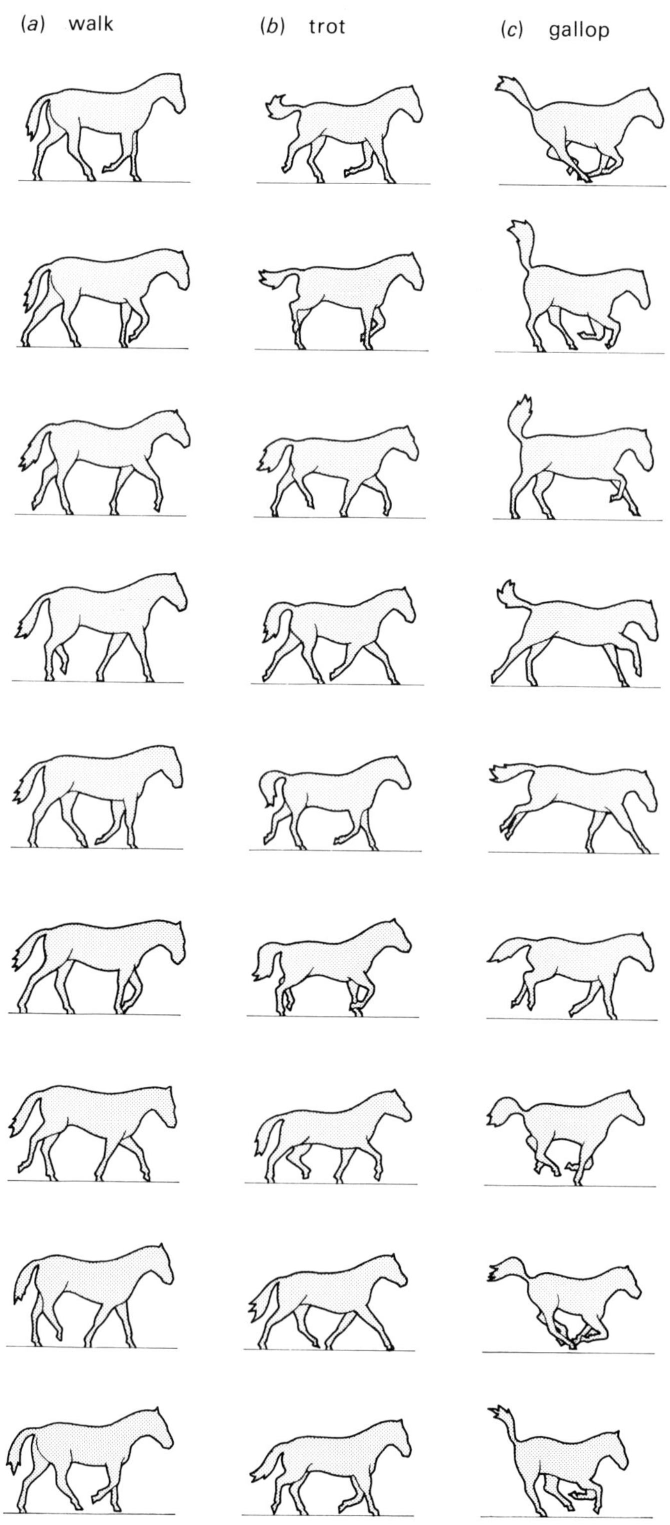

(a) walk   (b) trot   (c) gallop

SAQ 88 Turn back to figure 65 and describe how this elongation takes place in the horse.

SAQ 89 A rabbit can also move fairly rapidly. Is there evidence of distal elongation in its forelimbs? If so, say which bones are elongated.

SAQ 90 Consider also figure 66. What bones are in contact with the ground in
(*a*) man,
(*b*) horse,
(*c*) rabbit forelimb?
Explain how these differences can affect speed of motion in these animals.

Quadrupeds that walk but seldom run, like the pandas and bears, have feet similar to man and are said to be **plantigrade** (sole walking). Dogs, cats and others that move on their toes are described as **digitigrade**. The hooved mammals which touch the ground only with their enlarged claws are called **unguligrade**.

The speed at which muscles can contract to move limbs is limited, but the speed of a leg can be increased if different muscles can move different joints of the leg in the same direction at the same time. The total motion they produce will be greater than the motion of any one muscle working alone. The change from plantigrade to digitigrade or unguligrade gives the 'running' leg an extra limb joint which also thus increases rate of movement.

4.4.2 Walking and the arthropods

The tubular exoskeleton of an arthropod gives great strength coupled with lightness; it resists deformation and is suitable for the formation of levers. Because the skeleton is external, all the muscles are attached to its internal surfaces. Figure 108 shows a typical but simple arrangement of antagonistic muscles controlling joint position and limb movement in an arthropod.

Figure 109 shows the leg of a grasshopper which is fairly representative of an arthropod walking appendage. Its jointed nature can be clearly seen. Many of these joints are of the 'peg-and-socket' variety and permit movement in one plane only, as

108 Antagonistic muscles in an arthropod limb

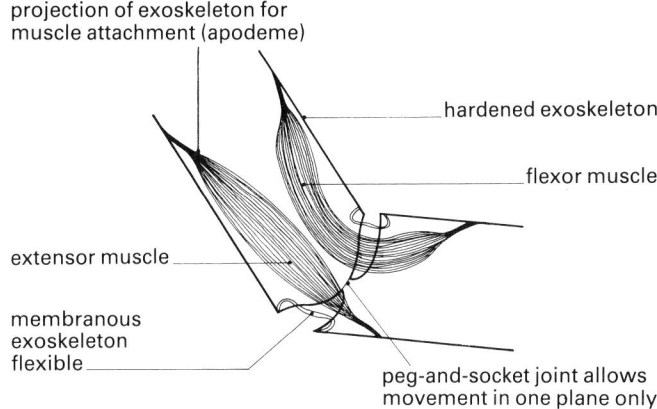

projection of exoskeleton for muscle attachment (apodeme)

hardened exoskeleton

flexor muscle

extensor muscle

membranous exoskeleton flexible

peg-and-socket joint allows movement in one plane only

109 Leg of a grasshopper

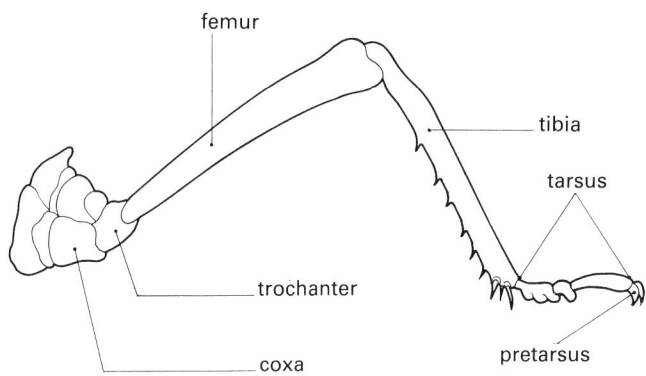

femur

tibia

tarsus

trochanter

pretarsus

coxa

does a hinge joint such as the human elbow. Different joints in the same limb will operate in different planes which increase the flexibility of their movement.

While an insect is walking, three legs are always in contact with the ground, maintaining a stable tripod. On one side, the first leg pulls and the third pushes while on the opposite side, the second leg acts as a prop. Meanwhile, the remaining three legs move forwards to repeat the process. Because the thorax does not twist, the effect is to swing the whole body from side to side in a rolling gait. Slow motion film reveals that legs do not move precisely together but have an off-beat rhythm.

4.5 Movement in water

Two problems face organisms that move in water. Water is almost 800 times as dense as air and therefore offers considerable resistance to movement unless the body of a swimming organism is specially adapted. This density also increases pressure on the body of organisms and they face problems if they are to penetrate deeper areas. Pressure increases by 100 g m^{-2} for every 10 m depth. However, this same density of water frees an organism from most of the force of gravity and makes support much easier.

This section will concentrate on the adaptations of fish for locomotion, studying streamlining, the organisation of muscles and skeleton, and the role of fins and buoyancy.

The mackerel in figure 110 shows the two important features of a fish which are associated with propulsion. These are the streamlined, cigar-shaped body, enabling the fish to move through the water with a minimum of resistance and the vertically held tail (caudal) fin.

110   A mackerel (*Scombus scombus*)

Figure 111 shows the effects of moving models of various shapes through the water. The resistance from water pushed aside in front is much less than the drag of eddying water behind. This explains why the widest point of a fish body is towards the front. Because a fish must actively propel itself, muscle arrangements dictate that the body is slightly compressed or oval in cross-section.

111   Movement of shapes through water

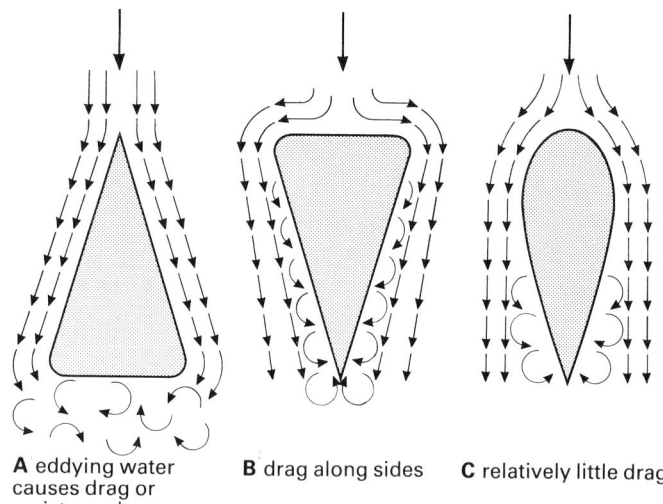

A eddying water causes drag or resistance here   **B** drag along sides   **C** relatively little drag

Figure 112 shows a sequence of stills from a cine-film of a moving fish. Examine it carefully.

112   Sequence of swimming movements of a fish

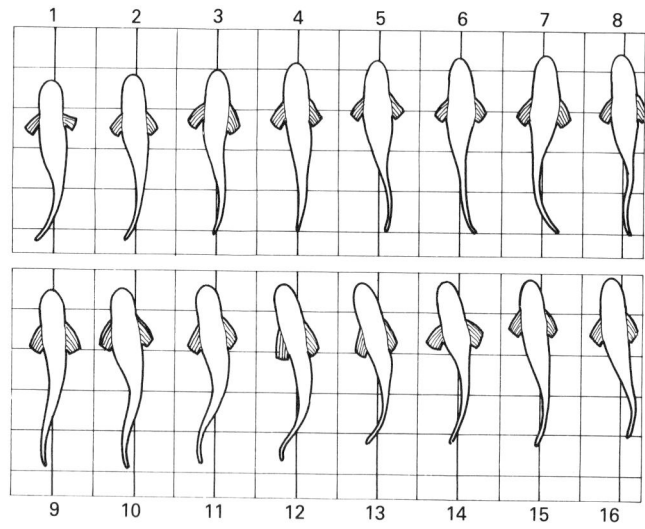

SAQ 91 Describe the movements of the tail as shown in figure 112. Use frame numbers for reference.

The movements of the tail described in your answer to SAQ 91 produce a backward and sideways force on the water. This propels the fish forwards.

4.5.1 The muscles of swimming

Movement of the tail is brought about by contraction of muscle blocks known as **myotomes**. The arrangement of myotomes in a dogfish is shown in figure 113. The muscle fibres in the myotomes are attached to sheets of connective tissue instead of bone. Contraction of muscle in one side of the body begins at the head end. A wave of contraction then passes down the body to the tail. A similar wave of contraction starts on the opposite side about half a second after the first and passes down to the tail.

113 Muscles associated with locomotion in a dogfish

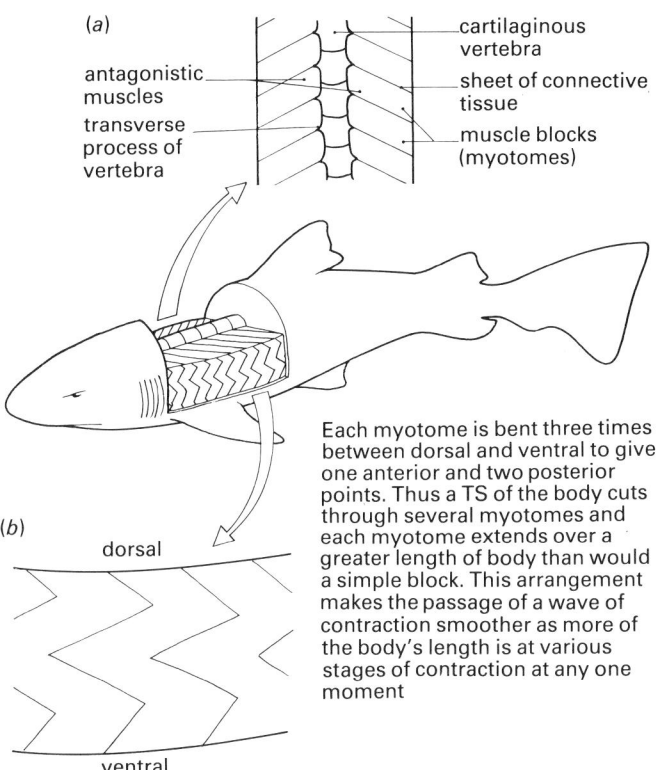

(a)
antagonistic muscles
transverse process of vertebra
cartilaginous vertebra
sheet of connective tissue
muscle blocks (myotomes)

Each myotome is bent three times between dorsal and ventral to give one anterior and two posterior points. Thus a TS of the body cuts through several myotomes and each myotome extends over a greater length of body than would a simple block. This arrangement makes the passage of a wave of contraction smoother as more of the body's length is at various stages of contraction at any one moment

(b)
dorsal
ventral

This is succeeded by a second wave of contraction on the original side, and so on.

SAQ 92 Make a sketch based on an extended figure 113(a) showing a wave of contraction beginning on both sides of the body. Indicate on your sketch the shape of the body and which muscles are contracted.

4.5.2 Stability in fish

The sideways force of the tail on the water might be expected to make the head end move from side to side. Such a movement is described as **yawing**. This is prevented in two ways.

Firstly, the head end is larger and has a greater mass than the tail. This tends to reduce sideways movement. Secondly, the dorsal and ventral vertical fins present a large surface area to resist sideways movement of the body. Also, many fish are flattened laterally. Both these effects reduce yawing.

In addition to yawing, the stability of a fish may be upset by **pitching** and **rolling** (see figure 114). The dorsal and ventral fins are vertical. The pectoral and pelvic fins are more or less horizontal.

114 Types of instability in a fish

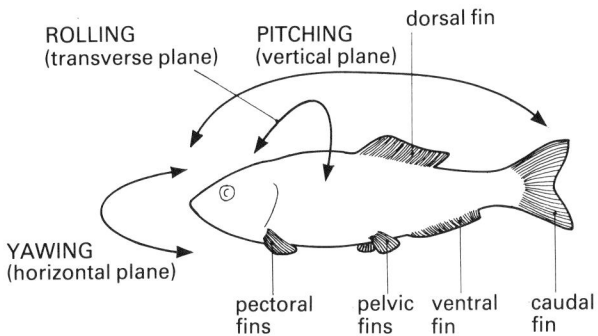

ROLLING (transverse plane)
PITCHING (vertical plane)
dorsal fin
YAWING (horizontal plane)
pectoral fins
pelvic fins
ventral fin
caudal fin

SAQ 93 Which fins would you expect to help counteract (a) pitching, (b) rolling?

4.5.3 Buoyancy in fish

In order to navigate through the water, a fish must be able to control its depth. Bone and muscle are denser than sea water and so tend to drag an animal down. Bony fish, like herring, are able to regulate their density by means of a **swim-bladder**.

In some fish, such as the goldfish, the swim-bladder is connected to the pharynx by an air duct. Thus, a goldfish changes its depth by swimming and then maintains its level in the water by swallowing air or 'blowing bubbles' via the mouth. Goldfish also have a special gas gland with a capillary network, and the swim-bladder may be recharged from gas already dissolved in the blood.

The more advanced teleosts, such as cod, have a closed swim-bladder. Gases are secreted into or absorbed from it by gas glands in the wall, aided by a special capillary area controlled by circular and radial muscles. A countercurrent system associated with fine capillaries enables gas under low pressure in the blood to enter the high pressure area of the swim-bladder. Figure 115 shows how most fish are able to float at any depth. Factors that confer buoyancy counteract the weight of the skeleton and muscles.

115 Maintaining buoyancy

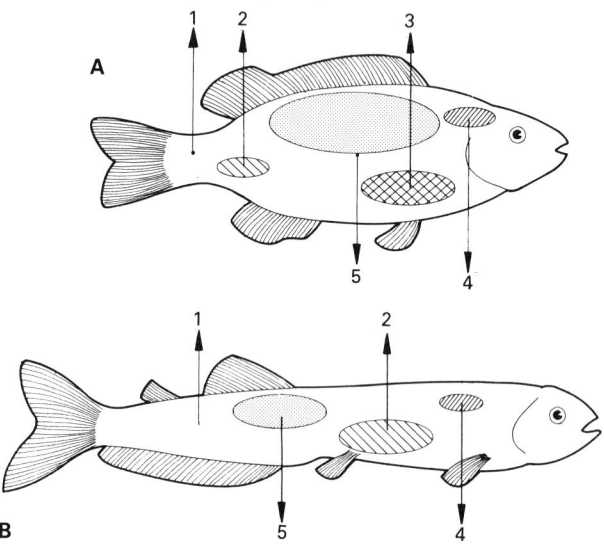

Most fish are able to float at any required depth. In fish with swimbladders (**A**), components that float are the body fluids (1), fat (2) and gas in the swimbladder (3). The tendency to float is counteracted by the weight of, for example, the skeleton (4) and the protein of the body (5). In fish without swimbladders (**B**), only the body fluids (1) and fats (2) give buoyancy, and these counteract the weight of the skeleton (4) and body proteins (5).

The cartilaginous fish, such as sharks, which have no swim-bladder, are able to control their depth in water due to the design of the tail-fin which is more developed on the ventral surface than the dorsal surface (see figure 116(*a*)). This **heterocercal** tailfin generates some upthrust as well as propelling the fish forwards through the water. The main lift comes from the larger pectoral fins, which work like aeroplane wings or hydrofoils. The large oily livers of these fish effectively reduce their density.

SAQ 94 What will happen to a shark when it stops swimming?

Bony fish have a **homocercal** tail (figure 116(*b*)) which contributes a considerable thrust forwards and acts as a rudder but does not generate upthrust.

116 Types of tail fin: (a) heterocercal tail of a cartilaginous fish; (b) homocercal tail of a bony fish

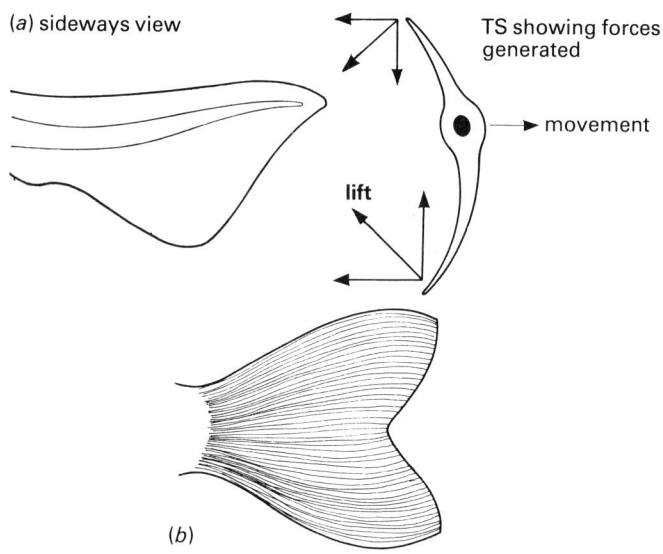

The pectoral fins of cartilaginous fish are large and set at a slight angle to the body. The pelvic fins are smaller but are set at an angle. They can be seen in figure 113.

As the fish moves through the water, these fins exert a downward and forward force on the water. This is counteracted by an upward and backward force by the water on the fish. This tends to lift the fish up in the water. The backward drag slows forward movement to some extent, but it is much less than the forward thrust of the tail.

4.5.4 Extension: Swimming in fish and other animals

Read chapter 2 of *'How animals move'* by James Gray.

Answer the following questions.

1 Explain how different animals have exploited the

mechanism of the sail, the jet, the paddle and the screw to propel themselves through the water.

2 State how swimming in an eel differs from swimming in a mackerel.

3 Discuss the role of streamlining, the ear and the lateral-line systems in locomotion of a fish.

4 Most fish use their tails for propulsion. Give examples of fish which use other methods.

4.6 Movement in air

There are three groups of animals in which flight is well developed. These are the birds, bats and insects. In each of these groups, flight is brought about by possession of wings.

SAQ 95 Suggest (*a*) an advantage which flying animals might have over non-flying animals. (*b*) Which sense organs might have undergone special development in flying animals?

Because air has a low density, it creates two major problems for a flying organism. First, air provides very little buoyancy, and second, still air is not a good medium against which to develop a thrust. In fact, air can only exert a force and provide support when there is movement between the wing and the air, either by the wing moving through the air or by air moving over the wing.

4.6.1 Bird flight

There are two types of flight: these are passive **gliding flight** and active **flapping flight**. Gliding will be considered first.

Wings are not set exactly horizontally. The leading or upstream edge is raised slightly in relation to the downstream edge. As air moves over the surface of the wings (or as the wings move through the air), two forces are exerted on them:

(*a*) a *lift force* which tends to raise the bird, and (*b*) a *drag force* which tends to move it backwards.

The resultant aerodynamic force is shown in figure 117.

117 Forces operating on a bird's wing during gliding

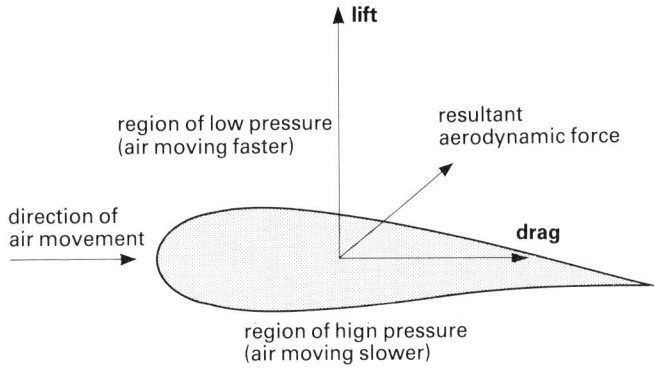

The downward force of gravity will also have an effect on flight. In practical terms, the ability to fly depends on the balance between the lift and drag forces and the pull of gravity.

SAQ 96 What factors might you expect to affect the lift effects in flight?

The speed of a glide depends on the weight of the bird and the size of the wings. A heavy bird with small wings glides rapidly. A light bird with large wings glides slowly.

Many birds use upward air currents to help keep them up in the air. Swifts can be seen gliding along the eaves on the windward side of a building. Gulls use upcurrents produced along cliffs by winds blowing onshore. In temperate and tropical countries, thermals, which are currents of warm air, rise from the ground in the morning. Birds such as buzzards, condors and vultures use them for gliding.

In the second type of flight, flapping flight, the bird's wings beat up and down to provide a lift **equal** to the weight of the bird and forward thrust **equal** to the backwards drag of the air. In this case, the wings are more comparable to the rotor blades of a helicopter than the wings of an aeroplane. Air resistance gives upthrust on the extended wing, and this is transmitted from the wing to the coracoid and sternum as the wing moves downwards. To rise or change speed the lift and thrust forces would momentarily have to be greater than the weight of the bird and the drag of the air.

Figure 118 shows the rate of wing beats for four different birds.

118 Rate of wing beat of different birds

	Mass in kg	Wing beats per s
heron	1.45	2
pigeon	0.333	7
wren	0.010	12
hummingbird	0.005	35

SAQ 97 Using figure 118 what is the relationship between size of bird and rate of wingbeats?

Flapping flight requires very strong wing muscles and large amounts of energy. Figure 119 shows the main bones and muscles associated with flight.

A bird skeleton is quite a rigid structure providing a stable framework for muscle attachment. There is considerable fusion of bones which contributes both to reducing weight and increasing rigidity.

The levator muscle passes through a groove where the scapula, coracoid and clavicle join, and is inserted on the upper side of the humerus. The depressor muscle arises from the sternum and clavicle (omitted from figure 119 for simplicity) and is attached to the underside of the humerus.

SAQ 98(*a*) Comment on the size of the keel. What do you think its function is?
(*b*) Name a structure resulting from fusion of bones.

SAQ 99 Which muscle is associated with (*a*) the upstroke and (*b*) the downstroke of the wing?

Studies on flight have shown that there are at least two types of wing movement. Those that occur during and immediately after take-off (slow flapping flight) are very energetic. In rapid flapping flight the upstroke makes little energy demand.

Figure 120 shows the sequence of movement in the flight of the Canada goose.

SAQ 100 Using figure 120, explain how and where lift and forward propulsion are produced.

As figure 120 shows, the primary feathers play a role in producing lift and propulsion. As they move upwards, the weight of the bird is supported on the inner wing. If compared to the mechanism of an aeroplane, the primaries can be said to act as propellers and the inner wings as the wings.

119 (a) Bird skeleton

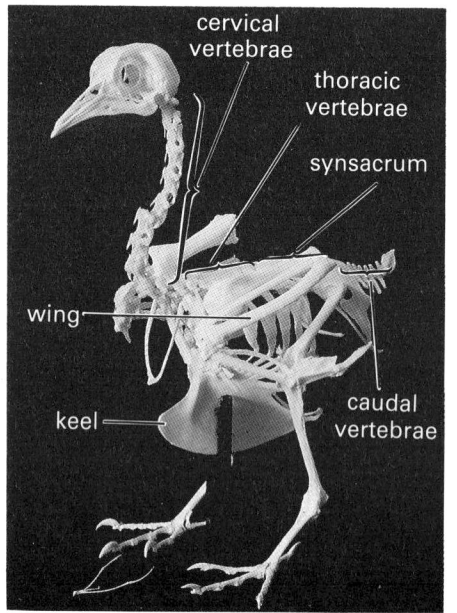

(b) flight apparatus

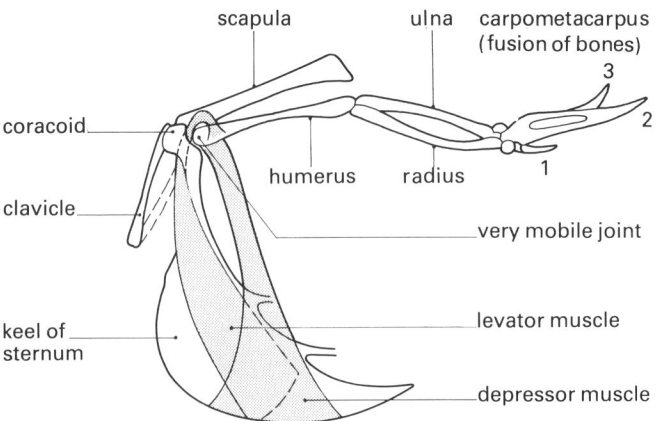

120 Flight movements of the Canada goose

(a) Downstroke

  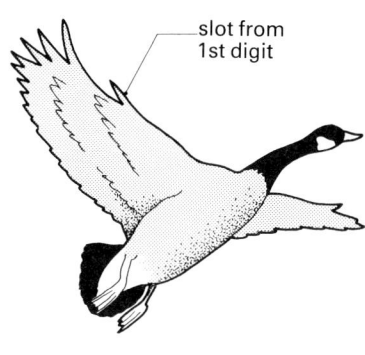

slot from 1st digit

(1) Wings extended and behind vertebral column

(2) Wings move downwards towards horizontal, lift is produced by the movement of the wing at an angle to the flow of air and forward thrust is also generated

(3) Primary feathers are bent upwards by upthrust of air as the wing descends, the 1st digit provides a feather 'slot' maintaining smooth air flow over the wing surface

(b) Upstroke

(6)

(4) Wings sweep forward in front of the body (free movement at the scapular joint)

(5) Wings parallel to each other in front of the body and upstroke begins with a sharp bending of the wrist followed by a flick which straightens the wrist again and brings the wings to their first position (1). Primaries separate and are bent upwards again by the upthrust and each separated feather generates some lift. Wings sweep back for new cycle

Practical L: Flight in birds

Wings and feathers are both characteristic of birds. Many aspects of their structure act as important adaptations for flight. In this practical you will examine both and consider their adaptations.

Materials

Whole mounted bird, primary feather, magnifying lens, monocular microscope and lamp, figure 63 (pentadactyl limb) from this unit

121 Structure of a bird's wing

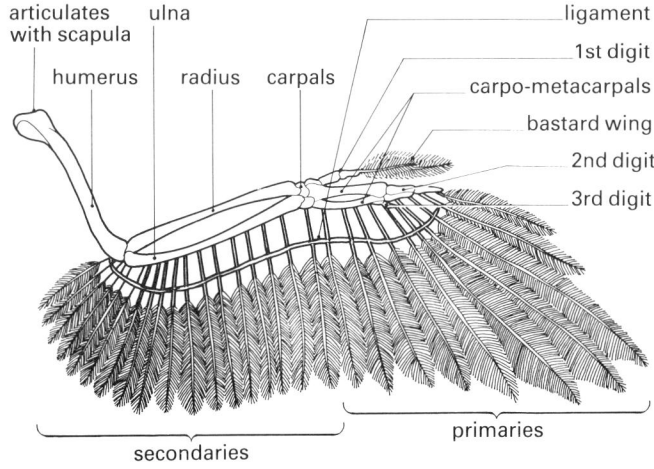

articulates with scapula — ulna — humerus — radius — carpals — ligament — 1st digit — carpo-metacarpals — bastard wing — 2nd digit — 3rd digit — secondaries — primaries

Procedure

(a) Identify the parts of the wing using figure 121

(b) Compare the wing with the basic pentadactyl limb illustrated in figure 63 of this unit.

(c) Examine the feather with the naked eye, the magnifying lens and the microscope.

(d) Notice the mass of the wing and the feather.

(e) Make drawings of your observations.

Discussion of results

1 What has happened to most of the carpals and metacarpals in a wing compared with the basic pentadactyl plan?

2 What has happened to the second carpo-metacarpal?

3 Which feathers are mainly associated with providing lift during the upstroke?

4 What implications do the mass of the wings and feathers have for flight?

5 What aspects of the feathers act as useful adaptations for flight? Consider such points as shape, position and the role of the tiny hooks.

6 Compare the structure of a bird's wing with that of a bat (figure 65).

7 Relate the flight of a bird to that of a flying fox (a fruit-eating bat).

8 Do any of the other body feathers have a direct effect upon flight?

Show this work to your tutor.

4.6.2 Insect flight

Of the invertebrates, the insects are the only group to have developed the ability to fly. The forces which keep an insect up in the air and move it forwards are produced by the movements of wings. Gliding flight is found in only a few insects, such as the Red

Admiral butterfly. The muscle action necessary to operate the wings requires much energy.

The mechanics of flight are fairly similar for all birds, but this is not the situation for insects. The insect exoskeleton has made possible a variety of relationships between the wings and the muscles that move them and this is reflected in the range of ways in which the wings actually move. However, general principles about wing movement can be stated.

(a) The wing moves in a continous sweep and, if the insect is held stationary, the tip of the wing traces a figure eight due to a combination of an up-and-down movement and a twisting at the wing base.

(b) In forward horizontal flight the wing vibrates in a plane oblique to the long axis of the body. This is shown in figure 122.

122   Plane of wing vibration in forward flight

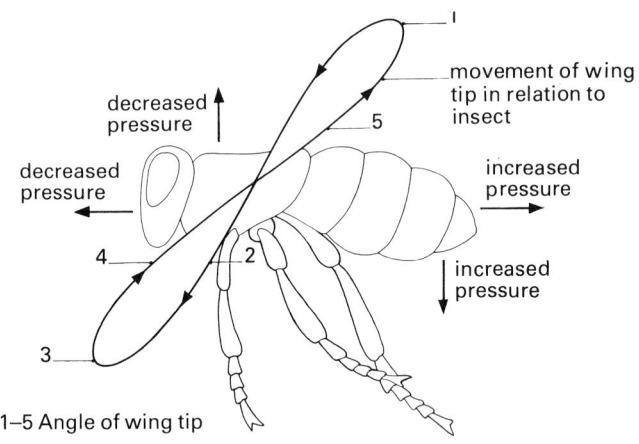

1–5 Angle of wing tip

(c) The propulsive force comes from the downward stroke with the anterior wing-edge held low and the posterior edge high. The upward stroke reverses this relationship. Lift is obtained on both the upstroke and the downstroke.

(d) These movements result in a zone of increased pressure in front and above. Thus, the insect moves forward along a pressure gradient.

The wing can function efficiently as an aerofoil as long as the airstream flowing towards it is free of turbulence. This would cause no problems for the anterior pair of wings, but their movement would create turbulence for the hind pair. Thus, flight is most efficient with two wings (that is a single pair). Many insects have become functionally two-winged (see figure 123).

123   Insect wings: (*a*) crane-fly; (*b*) bee

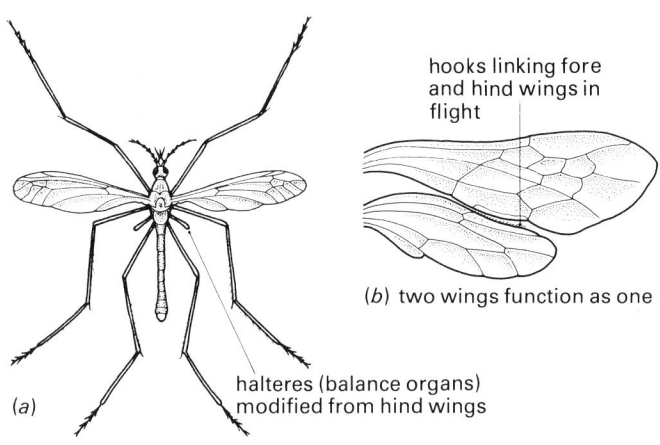

hooks linking fore and hind wings in flight

(*b*) two wings function as one

halteres (balance organs) modified from hind wings

(*a*)

SAQ 101 How have (*a*) a crane-fly, and (*b*) a bee become functionally two-winged?

SAQ 102 To what order of insects does the crane-fly belong? Name two other insects belonging to this order.

Among the beetles, the front pair of wings, the elytra, are stiff and held at an acute angle during flight. Locomotion is the role of the posterior wing pair. The dragon-flies use both pairs of wings for flight but they do not beat in phase. The posterior pair meets the backward stream of air before it has been displaced by the anterior pair.

Both pairs of wings possessed by an insect articulate with the thorax and each wing is composed of an extension of its outer layer, the **integument**. Wings are supported by a series of ribs and veins which are thickened hollow tubes. One of these supports the leading edge and the tip, and there is a tendency for veins to be thicker and closer together towards the leading edge, giving additional stiffness as in an aircraft wing. The trailing edge of an aircraft wing is thin and has hinged portions, flaps and ailerons which are used to control speed and stability in flight. A similar design is seen in the lobes of the insect wing, the shape and flexibility of which varies widely through the class Insecta. These lobes bend automatically as the wing moves and thus, their efficient action is a product of design rather than control. The pleating of some wings gives additional strength, rather in the manner of corrugated iron.

Movement of the wings of insects works on a quite different principle from that of birds. The flight muscles are attached not to the wings but to the body wall. Figure 124 shows how the action of the muscles moves the wings up and down in advanced insects such as flies, bees and beetles.

The wings are attached to the top (**tergum**) and sides (**pleura**) of the insect's thorax. They act as levers and are moved up and down by changes in shape of the thorax caused by contraction of two main sets of muscles. The vertical muscles run between the roof and floor of the thorax. The longitudinal muscles run between the anterior and posterior ends of the thorax.

124   Wing movements in insects

TS thorax

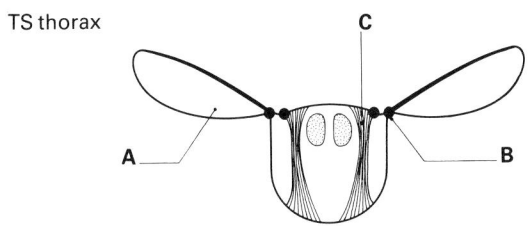

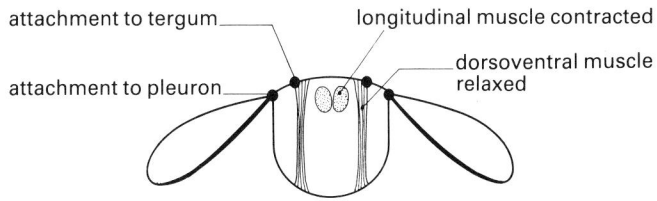

attachment to tergum

attachment to pleuron

longitudinal muscle contracted

dorsoventral muscle relaxed

SAQ 103 In figure 124, which of the points labelled **A**, **B** and **C** represent the fulcrum, effort and load?

SAQ 104 Compare the distance moved by the effort and the load during wing movement.

SAQ 105 How are insect flight muscles supplied with oxygen?

The distortion of the thoracic exoskeleton is eventually converted into wing movements by a complex system of cuticular levers near the wing base. Variations here create the variety of insect flight movements. More primitive insects, such as locusts and dragon-flies depend on muscles which run directly to the sclerites at the base of each wing.

4.7 Summary assignment 6

Use your answers to self test 5 as a summary of this section.

Self test 5, page 114, covers section 4 of this unit.

4.8 Extension: 'Muscle, a remarkable machine'

Ask your tutor for a copy of the article 'Muscle, a remarkable machine' by G. Goldspink in the *School Science Review* of September 1981.

This article reviews many of the topics studied in connection with movement.

4.9 Past examination questions

1 Figure 125 is a drawing of the forelimb skeleton and part of the shoulder girdle of a mammal (rabbit).

(*a*) Copy the figure and complete it to show the following muscles and their attachments to the skeleton.

  (i) A muscle which contracts in preparation for landing after a leap. Label this muscle **X**.

  (ii) A muscle which will extend the digits. Label this muscle **Y**.

(iii) A muscle which will contract in response to a painful stimulus applied to the forefoot. Label this muscle **Z**. (6 marks)

125 Forelimb and girdle of a rabbit

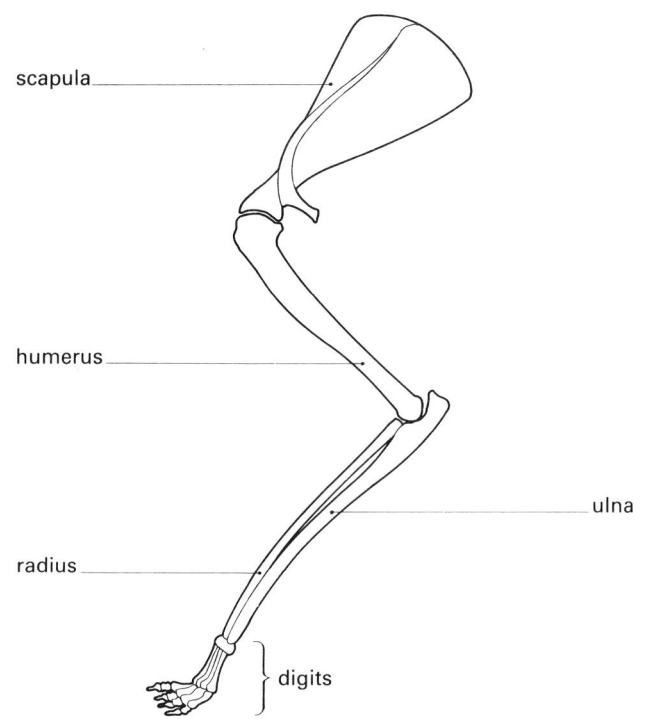

(*b*) Make a labelled diagram of the reflex arc which would be responsible for the action in (*a*) (iii).
(10 marks)

(London, 1984)

2 Describe the propulsive forces which bring about movement in a *named* aquatic animal. (9 marks)

Show how the animal (*a*) turns to the left and (*b*) rises. (4,4 marks)

To what extent can the movements of an aerial animal be regarded as operating on similar principles? (8 marks)
(London, 1979)

3 Give an illustrated account of the way in which skeletal and muscle systems operate together in a *named* terrestrial mammal to produce rapid locomotion. Briefly contrast this organisation with that of an insect leg.
(Oxford & Cambridge, 1983)

4.10 Recommended reading for sections 1–4

Physiology of mammals and other vertebrates by P.T. Marshall & G.M. Hughes, 2nd ed., chapters 8 and 9.

How invertebrates live by Kaye Mash, chapter on 'Skeletons and movement'.

How fishes live by Peter Whitehead, chapter on 'Going places'.

Institute of Biology, Studies in Biology Series:
11, *Muscle* by D.R. Wilkie, 2nd ed.
22, *Animal skeletons* by J.D. Currey.
33, *Animal flight* by C.J. Pennycuik.

Insect flight by J.W.S. Pringle.

Muscle contraction by W.F. Harrington.

The contractile behaviour of mammalian skeletal muscle by A.J. Buller & N.P. Buller.

Cilia and related organelles by P. Satir.

Buoyancy of marine animals by E.J. Denton.

How we control the contraction of our muscles by P.A. Merton, in *Scientific American*, **226**, no.5, pp.30–7, May 1972, and also as an offprint, no.1249.

Animal locomotion by James Gray.

Locomotion of animals by R. McNeill Alexander.

On size and life by T.A. McMahon & J.T. Bonne.

This list deals exclusively with animals. Many general texts on plants deal also with support and movement.

Section 5 Behaviour

5.1 Introduction and objectives

All organisms change in response to changes in their environment. The capacity to respond to stimuli, **irritability** or **sensitivity**, is an essential property of life. Unit 6, *Response to the environment*, studied a variety of these responses and the biological systems on which they are based. Responses that promote survival and are reversible can be termed **behaviour**. Although tropic movements in plants are adaptive responses, they are brought about by growth and are not reversible. In general, the responses of plants are not considered as behaviour, though there are exceptions. The trapping activities of the Venus fly-trap which were examined in ABAL units 1 and 6 (*Inquiry and investigation* and *Response to the environment*) can certainly be considered as 'behaviour'. However, because 'behaviour' is applied to reversible movements of the whole or part of an organism, this branch of biology is largely concerned with the activities of animals and their reactions to their environment.

Animals have solved the problems to which they are exposed in a variety of ways and, as one ethologist has said, 'understanding the diversity of the animal kingdom requires that the behaviour of each species be seen in relation to the environmental context to which it has been adapted'. **Ethology** is the approach to the study of behaviour which holds fundamental that animals must be studied, at least in part, in their natural conditions if useful questions about their behaviour are to be asked and answered. Some comparative psychologists have used methods where animals are isolated and all possible variables controlled, but it is now recognised that the value of such findings is limited by the unnatural environment. An extreme position is taken by those who declare that behaviour is nothing but neurophysiology and sensory physiology (the function of nerve cells and receptor organs), while some biochemists believe that learning can be explained in terms of molecules. What is clear is that many different areas of scientific research can contribute to the understanding of behaviour.

Young herring gulls ask for food by pecking at their parents' bills. Scientists observing this behaviour might approach its study in four different ways, asking four different kinds of questions.

– What stimuli and mechanisms are involved? (What causes the behaviour?)

– How does the behaviour develop in the chick? (Is it present immediately on hatching? How long does it continue? And so on...)

– How did this behaviour evolve? (Do closely related birds show a similar reaction? How is it inherited?)

– What is the advantage of possessing this piece of behaviour? (What is its survival value? Why is it sustained by natural selection?)

These four kinds of question may be termed immediate causation, development, evolution and function, and in this section you will encounter examples of behaviour from all four viewpoints.

Because behaviour is such a wide and complex subject, this section is intended only to serve as an introduction to a topic which integrates many of the areas studied in biology courses and leads to an understanding of the functioning of the whole animal within its environment.

After completing this section you should be able to do the following.

(*a*) List the ways in which behaviour contributes to the survival of individuals and species.

(*b*) Suggest areas that should be investigated when studying behaviour.

(*c*) Compare the advantages and disadvantages of field and laboratory experiments in studying behaviour.

(d) List the techniques and equipment that can be utilised by a behavioural scientist.

(e) Describe the use of models in behavioural investigations with named examples.

(f) Define and explain the importance of sign stimuli or releasers in relation to aggressive behaviour in robins and feeding in herring gulls.

(g) Explain the terms innate and learned behaviour and evaluate the usefulness of these terms.

(h) Explain and recognise examples of the following types of behaviour: reflex behaviour, kinesis, taxis, fixed action pattern, habituation, classical conditioning, operant or instrumental conditioning, trial-and-error learning, latent learning, imprinting, insight learning or reasoning.

(i) Define motivation and explain the factors that may be involved with it.

(j) Discuss the effect of hormones on behaviour with particular reference to sexual behaviour.

(k) Describe how hormones interact with other stimuli, as in the reproductive behaviour of the ring dove.

(l) Explain how biological rhythms may influence behaviour.

(m) State the effects of photoperiodism in animals and state how they may be mediated.

(n) List the ways in which animals may communicate with special reference to communication in honey-bees, pheromones and song.

(o) Discuss courtship as a means of communication.

(p) State the advantages of social life for animals and list the distinguishing characteristics of animal societies.

(q) Indicate the approaches used in the study of the evolution of behaviour and give evidence for the evolution of behaviour in lovebirds.

(r) Give an example of the effect a gene may have on behaviour.

5.2 The importance of behaviour in mammals

Behaviour is what an animal does and it cannot be isolated from other aspects of an animal's life. Nutrients required for a balanced diet would not be obtained without the ability to search for and select food. Many animals could not reproduce without their special behaviour patterns of courtship, mating and rearing of young. The characteristic animal populations found during an ecological investigation of any particular habitat would not exist without the behavioural responses of animals to particular environmental conditions such as humidity, vegetation, temperature and light availability. The behaviour patterns of an animal may aid or hinder its struggle for survival.

AV 1: Behaviour for survival

The video material shows examples of vertebrate behaviour patterns which contribute to the survival of the individuals concerned and/or to the survival of the species.

Materials
VCR and monitor
ABAL video sequence: *Behaviour for survival*
Worksheets

Procedure

(a) Check that you have all the relevant materials listed above.

(b) Check that the video-cassette is set up ready to show the appropriate sequence – *Behaviour for survival*.

(c) If you do not understand anything, then stop the sequence, rewind and study the relevant material again before consulting with your tutor.

(d) If possible, work through the sequence and worksheets with a small group and discuss the material with them. Explain the relevance of each behavioural activity to the life of the animal.

5.3 The study of behaviour

Careful observation, accompanied by detailed recording, is the basic tool in the study of behaviour. Perhaps the pioneer of behavioural study was a Frenchman, J. Henri Fabre, who originally trained as a teacher but lost his post in the 1870s because he included girls in his classes where he taught about the reproduction of plants and animals.

Charles Darwin, whose own work on animal behaviour paved the way for objective experiments and observations, called Fabre 'an incomparable observer'. Fabre's books were based on long and patient observation of insect behaviour and the minute and orderly records made of what he saw. The following is a quotation from *The study of instinct* published in English in 1918. It concerns the study of sexton beetles (*Necrophori*) shown in figure 126, which bury the bodies of small animals as food for their larvae. Their strong legs dig away earth from beneath the carcase until it is almost concealed in the hole and the females then lay their eggs. Here, Fabre records an experiment to test the belief of some that these beetles will move a body found on an unsuitable substrate and even summon help to do so.

126 Sexton beetle

'Yes, let us be simple without being childishly credulous. Before making insects reason, let us reason a little ourselves – let us, above all, consult the experimental test. A fact gathered at hazard, without criticism, cannot establish a law...

The Mouse so greatly desired is mine. I place her upon the centre of the brick. The grave-diggers under the wire cover are now seven in number, of whom three are females. All have gone to earth: some are inactive, close to the surface; the rest are busy in their crypts. The presence of the fresh corpse is promptly perceived. About seven o'clock in the morning, three Necrophori hurry up, two males and female. They slip under the Mouse, who moves in jerks, a sign of the efforts of the burying-party. An attempt is made to dig into the layer of sand which hides the brick, so that a bank of sand accumulates about the body.

For a couple of hours the jerks continue without results. I profit by the circumstance to investigate the manner in which the work is performed. The bare brick allows me to see what the excavated soil concealed from me. If it is necessary to move the body, the Beetle turns over; with his six claws he grips the hair of the dead animal, props himself upon his back and pushes, making a lever of his head and the tip of his abdomen. If digging is required, he resumes the normal position. So, turn and turn about, the sexton strives, now with his claws in the air, when it is a question of shifting the body or dragging it lower down; now with his feet on the ground, when it is necessary to deepen the grave.

The point at which the Mouse lies is finally recognised as unassailable. A male appears in the open. He explores the specimen, goes the round of it, scratches a little at random. He goes back; and immediately the body rocks. Is he advising his collaborators of what he has discovered? Is he arranging matters with a view to their establishing themselves elsewhere, on propitious soil?

The facts are far from confirming this idea. When he shakes the body, the others imitate him and push, but without combining their efforts in a given direction for, after advancing a little towards the edge of the brick, the burden goes back again, returning to the point of departure. In the absence of any concerted understanding, their efforts of leverage are wasted. Nearly three hours are occupied by oscillations which mutually annul one another. The Mouse does not cross the little sand-hill heaped about it by the rakes of the workers.'

(Further details of burying attempts follow with observations on the roles played by males and females. Finally, after six hours, the mouse is buried in softer soil.)

'It is absolute nonsense to speak of their first preparing the grave to which the body will afterwards be carted. To excavate the soil, our grave-diggers must feel the weight of their dead on their backs. They work only when stimulated by the contact of its fur. Never, never in this world do they venture to dig a grave unless the body to be buried already occupies the site of the cavity. This is absolutely confirmed by my two and half months and more of daily observations.'

His further discussion of the observations serves to dismiss any ideas of communication and cooperation.

'Were they not rather five chance Necrophori who, guided by the smell, without any previous understanding, hastened to the abandoned Mouse to exploit her on their own account? I incline to this opinion, the most likely of all in the absence of exact information.'

SAQ 106 What evidence is there, in this short extract, of Fabre's scientific approach to the study of behaviour? List three points or aspects of his approach that would make you consider his account to be reliable.

SAQ 107 State briefly what this quotation tells you about the behaviour of sexton beetles.

As the study of behaviour developed, two distinct schools of approach began to emerge. One, in Europe, was concerned with observing and testing animals under natural conditions, and the other,

based in America, was interested in learning about behaviour under controlled laboratory conditions.

Konrad Lorenz, 'the father of modern ethology', represents the natural approach which will be considered first. Lorenz laid the groundwork for many lines of research still being followed today. He was convinced that an animal's behaviour, as any physical adaptation, was part of its equipment for survival and the product of adaptive evolution. The following extract shows the kind of observations used by Lorenz and the way he reasons from these observations. Lorenz used a pair of greylag geese to hatch a Muscovy duck's eggs.

'From the seventh week of their life, however, the young Muscovies had nothing more to do with their former foster-parents nor with any other Greylag Geese, but behaved socially toward one another, as well as toward other members of their species, as a perfectly normal Muscovy Duck should do. Ten months later the one male bird among these young Muscovies began to display sexual reactions and, to our surprise, pursued Greylag Geese instead of Muscovy Ducks, striving to copulate with them, but he made no distinction between male and female geese.'

Then...

'In 1936, I kept a young Greylag isolated from its kind for over a week, so that I could be sure that its following-reaction was securely attached to human beings. I then transferred this young goose to the care of a Turkey hen, whom it soon learned to use as a brooding Kumpan for warmth instead of the electric apparatus it had hitherto favoured. The gosling then followed the Turkey hen, provided that I was not in sight, and kept this up for a fortnight; but even during that time I had only to walk near the two birds, to cause the gosling to abandon the hen and follow me. I did this but three times, to avoid conditioning the gosling to my person as a leader. When, after two weeks, the gosling began to become more independent of the warming function provided by the Turkey hen, it left her and hung around our front door, waiting for a human being to emerge

127 Konrad Lorenz in the company of imprinted geese

and trying to follow it when it did so. Now this gosling, excepting the few necessary trial runs, each of which did not last more than about a minute, had never actually consummated its following-reaction with a human for its object. On the other hand, for more than two weeks, it had been in constant contact with the Turkey hen; yet its following-reaction did not become conditioned to the Turkey in preference to the human. I would even suspect that its instinctive following-reaction

was never really released by the Turkey at all, and that its following the hen was predominantly a purposive act directed to the necessity of getting a warm-up from time to time. It never ran directly after the Turkey hen in the intensive way in which it would follow me and in which Greylag goslings follow their normal parents, but just kept near her in a most casual and deliberate sort of way, quite different from the normal reaction. Most impressive is the fact of the irreversibility of imprinting in such cases, in which birds become conditioned to an inaccessible object or to one with which it is physically impossible to perform the reaction.

The process of acquiring the object of a reaction is in very many cases completed long before the reaction itself has become established, as seen in the observations on the Muscovy drake cited above. This offers some difficulties to the assumption that the acquiring process in imprinting is essentially the same as in other cases of 'conditioning,' especially in associative learning.'

Here, Lorenz is examining ideas related to the phenomena of 'imprinting' in birds, an unusual or specialised form of learning which you will study further in section 5.4.9.

SAQ 108 In the example of the Muscovy ducks reared by greylag geese, what was the unexpected outcome?

SAQ 109 What similarities are there between the two examples quoted?

SAQ 110 What ideas about 'imprinting' can you gain from reading this account?

5.3.1 Equipment for improving observation

The kinds of observations made by Fabre and Lorenz can be supplemented but not replaced by a variety of aids, some very simple and others quite complex. Here is a list of some techniques that are used to supplement unaided observation.

Photography – *stills*
                cine-films
                time-lapse
                    (photographs taken at regular intervals, for example every 30 s, 5 min or 1 h)
                multiple-flash
                    (several flashes per single exposure. The flashes may be provided by a stroboscope. See figure 128.)

Video-filming
Tape-recording

SAQ 111 For each technique listed, consider how it might supplement and increase observation.

An observer studying the behaviour of small animals may wish to know if any patterns can be found in their periods of activity. For this, a record must be made of the activity of the animal over a period of several days for any patterns to emerge. Such work is tedious, and time-consuming and **activity recorders** of various kinds have been developed.

128 Multiple-flash photograph

For small rodents an activity wheel attached to a recording device can be used to detect wheel-running behaviour, as shown in figure 129.

129 An activity wheel

Studies on rats reveal a daily cycle of wheel-running and rest with less-marked rhythms every 2–4 h. Another way of detecting walking-type activities is to use a **photocell**. An infra-red light beam is shone across the cage (or tank in the case of aquatic organisms) so that the animal's movement interrupts the light beam and is thus detected and recorded. More than one beam or a series of mirrors can be used in conjunction to detect movement over smaller areas.

SAQ 112 Why is infra-red light used?

Tilt cages with pivoted floors and micro-switches at either end are another alternative, provided the slight floor movement does not influence behaviour.

Tape-recordings played back at high speed through a sound-level meter can also record animal activity periods.

For recording behaviour in the field, the microprocessor is just beginning to reveal its potential. Depending upon the keyboard skills of the user, behaviour can be recorded as frequently as once per second and the behaviour of several subjects can be recorded using a pre-defined code. It is possible to obtain a 'complete record' of behavioural events (in the laboratory or the field), which was previously only possible with film or video. The record obtained with this type of program can be stored on the built-in microcassette for transfer to a larger computer and later statistical analyses. A hard copy of the raw data can be produced by the built-in printer as soon as the observation finishes. Statistical analyses of quantitative results is becoming very important in some aspects of ethology.

5.3.2 The analysis of behaviour: Tinbergen and von Frisch

Although field observations give the most accurate view of behaviour, the diversity of stimuli present make it difficult to determine which are causing the behaviour. This can be overcome by carrying out laboratory experiments or, alternatively, by 'taking the laboratory into the field' and modifying the natural environment.

This is largely the approach adopted by Niko Tinbergen, one of the most influential ethologists, and for many years Professor at Oxford. He found that his long hours of field observation enabled him to learn about behaviour patterns and give him ideas for theories which he could later test by experiment. In his book *Social behaviour in vertebrates* he describes how the use of cardboard models led to an exact understanding of the stimuli which elicit the begging response of the herring gull chick, an **innate** reaction.

'We will now consider the begging response of the Herring gull chick more closely, because here we know exactly to what stimuli the chick responds.

It is possible to release the begging response of a newly born, inexperienced chick by presenting it with a flat cardboard model of the parent's head. The chick responds to this just as well as to the real head. The bill tip of the adult herring gull bears a red colour patch which stands out quite conspicuously against the yellow background of the bill itself. When this red patch is absent in a model, the chick will respond much less vigorously than to the normal model with the red patch (figure 130).

When these two models were presented in turn to a number of chicks, the average number of responses to the model without a red patch was only one-fourth of that to the normal model. Models in which there was a patch, but of colour other than red, released intermediate numbers of responses. This depended on the degree of contrast between the patch and the bill colour. In the same way, viz. by comparing the chick's

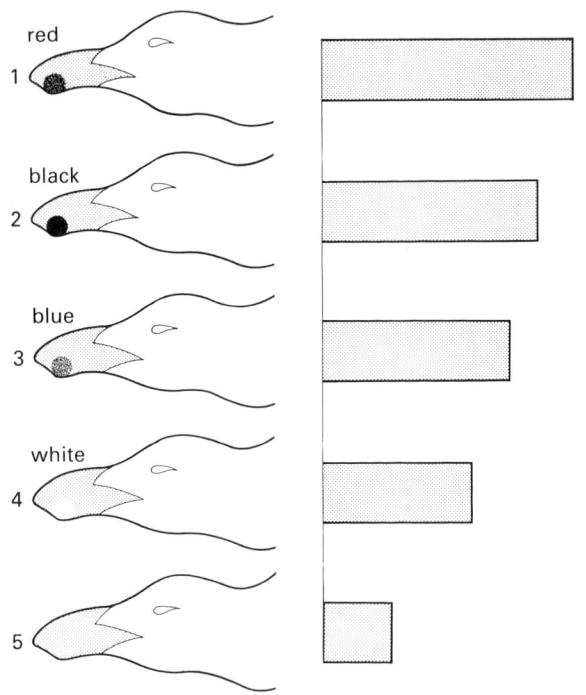

130 Models of herring gull heads used to release begging responses in newly born chicks. Colour of the mandible patch varied (1–4) or absent (5). Columns indicate the relative frequency of chick's responses

responses to various models, it was possible to study the influence of the yellow colour of the bill. Surprisingly enough, the colour of the bill in the models did not make the least difference to the chicks, except that a red bill released twice as many responses as any other colour (figure 131).

A bill in the natural yellow colour did not release more responses than did a white, a black, a green, or a blue bill. Neither did the colour of the head make any difference: one would expect that a

131 Releasing value of herring gull models with uniform bills of varying colour

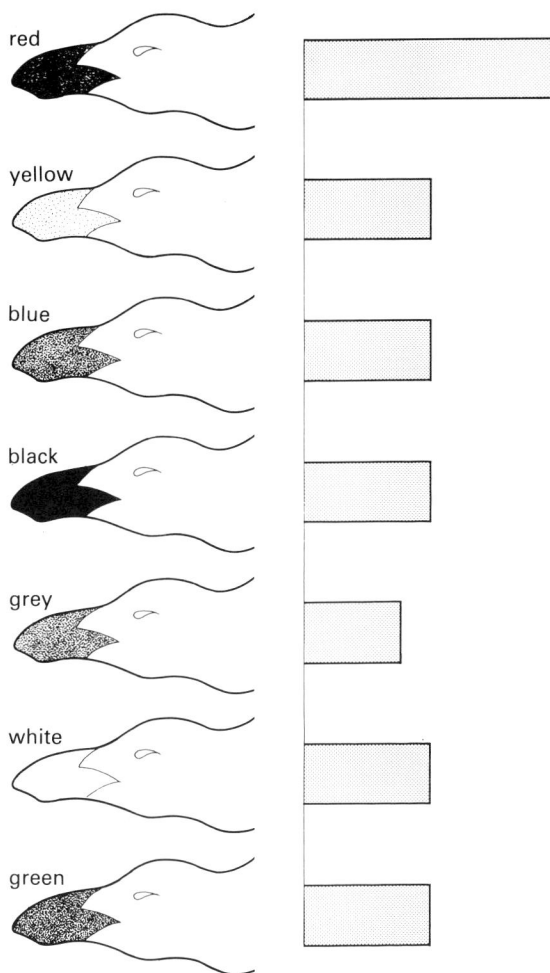

how they fulfil, so to speak, all the chick's expectations. The parent walks up to the chick, presents its bill in an almost vertical position, pointing the tip down, and it has a red blotch at the tip of the bill. This close correspondence between the characters of the parents and the stimuli to which the chick responds is amazing when we recall that the chick cannot 'know' what the parent looks like, or how it behaves.'

132 Female (left) and male American woodpecker

SAQ 113 Figure 132 shows male and female American woodpeckers. Male birds generally accept female birds but attack other males. A visual signal is believed to be involved in this case. Suggest what visual feature might be involved and describe a simple experiment using cardboard models that could be used in an attempt to test your hypothesis. Suggest an additional investigation using a female bird.

Karl von Frisch's contribution to the knowledge of animal behaviour lay particularly in an understanding of sensory capacities. He challenged the belief prevailing at the beginning of the century that bees were colour-blind by a series of simple experiments using coloured cards. He is best known for his work on communication in honey-bees. Like Tinbergen, many of his investigations were carried out in natural surroundings using a combination of observation and simple experimental techniques. The account which follows demonstrates the commonly used procedure of marking individual animals in some way so that they can be more easily identified.

white head would release more responses than a black or a green head, but that was not so. Nor did the shape of the head matter; it did not even make much difference when there was no head at all, but merely a bill. Yet the chicks can see the head very well, for they peck occasionally at the base of the bill, and even at the parent's red eyelids. When the chicks are hungry, there is just one thing to them that matters: the parent's bill with the red tip. In addition, the bill must be thin and elongate, it must point down, it must be as near the chick as possible, and as low as possible. But these are all the stimuli; everything else is irrelevant to the chick. It is remarkable how well the parent's behaviour and colour fit in with this,

'The following experiment shows how accurately the dancer's indication of distance is understood and followed by the other bees. Several bees individually numbered with coloured dots are, at a certain distance from the hive, fed with a sugar solution to which a scent, for example lavender oil, was added. Upon their return to the hive the numbered gatherers dance, and during the dance the hive companions smell the lavender and search for this specific odour when flying out. On the direct line from the hive to the feeding place and even further away 'scent plates' are placed at various distances. They emit the scent, but offer no food. The bees looking out for this odour at the indicated distance are attracted by the scent plates if they come near them. They fly around them and finally alight on them and are thereupon counted by an observer. In figure 133 we show the results of two such experiments. In the first trial the feeding place was at a distance of 750 m and in the second at 2000 m from the hive. The data given for different points of the graph indicate how many of the searching bees came to each scent plate during the time of observation. The graphs show that the indication of distance given by the dancer is quite accurate and has been well understood by the other bees.'

SAQ 114 What features of the natural environment have been altered?

SAQ 115 Why did the experimenters not include a food supply but just scent plates?

5.3.3 Laboratory behaviour: Pavlov, Skinner and Harlow

Ivan Pavlov's career was centred on the laboratory and the controlled experiment. His name is associated with the idea of the **conditioned response** (sometimes referred to as 'classical conditioning' or the 'conditioned reflex').

In his classical experiment, Pavlov used a dog which was lightly restrained in a harness. He repeatedly blew meat powder into its mouth and made accurate records of the amount of saliva it produced. (A fistula – a hole – was present in the cheek, from the

133 Distance indication in honey-bee communication. The distance of the scent plates from the hive are indicated on the x axis in metres. The numbers on the points of the graph correspond with the number of bees which have come to each scent plate. (a) Trial of June 27, 1949: distance between feeding place (F) and hive is 750 m, duration of experiment is 90 min. (b) Trial of July 20, 1952: distance to feeding place is 2000 m, duration of experiment is 3 h.

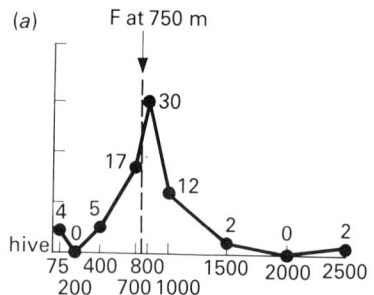

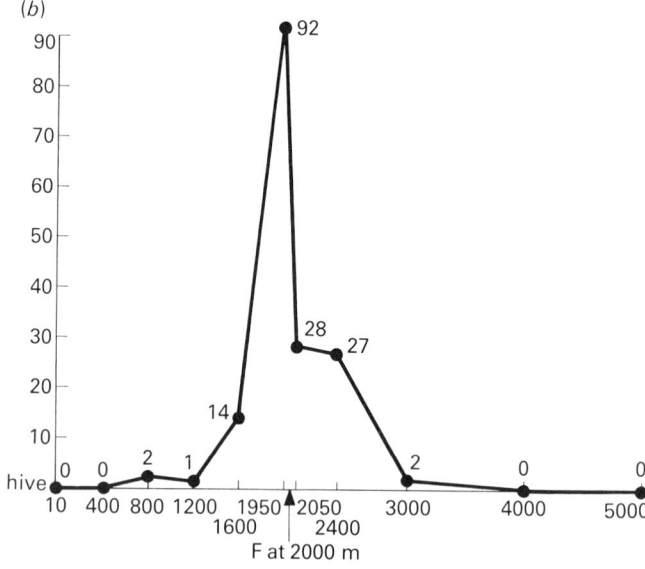

salivary duct and drops of saliva fell from a funnel and could be counted.) The external environment was carefully controlled. Pavlov discovered that a standard quantity of meat powder resulted in the secretion of a certain amount of saliva. This is thought to be a simple inborn reflex involving meat powder (stimulus), taste buds (the receptors), sensory, association and motor neurons and an effector which is the salivary gland. (Reflex actions are studied in the unit *Response to the environment*.)

Pavlov then started a metronome (an instrument which marks time at a selected rate by means of a pendulum) before puffing meat powder into the dog's mouth. At first, the stimulus produced no observable response except for momentary 'ear-pricking'. However, after about six demonstrations of the metronome ticking followed immediately by the meat powder, saliva was produced as soon as the metronome was started and *before* the meat powder arrived.

'We come now to consider the precise conditions under which new conditioned reflexes or new connections of nervous paths are established. The fundamental requisite is that any external stimulus which is to become the signal in a conditioned reflex must overlap in point of time with the action of an unconditioned stimulus. In the experiment which I chose as my example the unconditioned stimulus was food. Now if the intake of food by the animal takes place simultaneously with the action of a neutral stimulus which has been hitherto in no way related to food, the neutral stimulus readily acquires the property of eliciting the same reaction in the animal as would food itself. This was the case with the dog employed in our experiment with the metronome. On several occasions this animal had been stimulated by the sound of the metronome and immediately presented with food – i.e. a stimulus which was neutral of itself had been superimposed upon the action of the inborn alimentary reflex. We observed that, after several repetitions of the combined stimulation, the sounds from the metronome had acquired the property of stimulating salivary secretion and of evoking the motor reactions characteristic of the alimentary reflex.'

Pavlov believed that the study of conditioned reflexes would 'throw open the door to a true physiological investigation probably of all the higher nervous activities of the cerebral hemispheres'.

SAQ 116 Write a simple description of a conditioned reflex.

B.F. Skinner acknowledged the significance of Pavlovian conditioning in explaining certain responses of organisms, but believed that voluntary and active responses were more important than reflexes. He gave the name **operants** to the behaviour that operates or acts on the environment to produce consequences that, in turn, have an effect on the performer. The results of an organism's activity determine whether it performs a similar action or tries something else.

Skinner summarises his approach to the investigation of operant conditioning in a lecture given in 1956.

'In studying such behaviour we must make certain preliminary decisions. We begin by choosing an organism – one which we hope will be representative but which is first merely convenient. We must also choose a bit of behaviour – not for any intrinsic or dramatic interest it may have, but because it is easily observed, affects the environment in such a way that it can be easily recorded, and for reasons to be noted subsequently may be repeated many times without fatigue. Thirdly, we must select or construct an experimental space which can be well controlled.'

134   A Skinner box

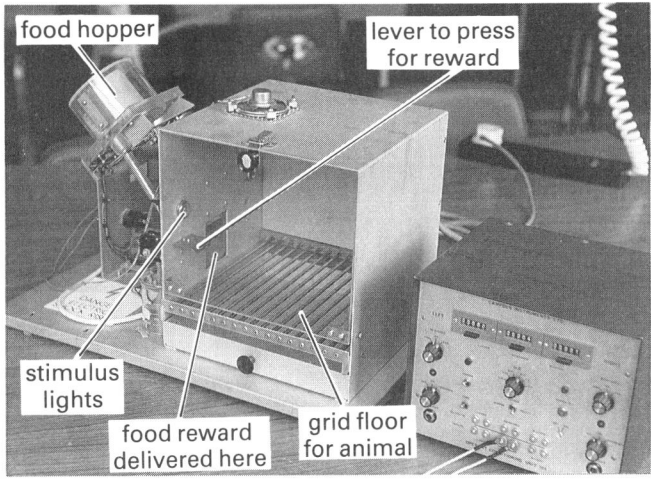

food hopper

lever to press for reward

stimulus lights

food reward delivered here

grid floor for animal

Skinner's requirements were met using an apparatus rather like the arrangement in figure 134. The partly sound-proofed box would contain a pigeon which could be, in Skinner's words, seen in the act of pecking a translucent plastic plate behind a circular opening in the partition. The plate is part of a delicate electric key; when it is pecked, a circuit is closed to operate recording and controlling equipment. Coloured lights can be projected on the back of the plate as stimuli. The box is ventilated, and illuminated by a dim ceiling light.

We are interested in the probability that in such a controlled space the organism we select will engage in the behaviour we thus record. At first blush, such an interest may seem trivial. We shall see, however, that the conditions which alter the probability, and the processes which unfold as that probability changes, are quite complex. Moreover, they have an immediate, important bearing on the behaviour of other organisms under other circumstances, including the organism called man in the everyday world of human affairs.

Probability of responding is a difficult datum. We may avoid controversial issues by turning at once to a practical measure, the *frequency* with which a response is emitted. The experimental situation (shown in figure 134) was designed to permit this frequency to vary over a wide range. In the experiments to be described here, stable rates are recorded which differ by a factor of about 600. In other experiments, rates have differed by as much as 2000:1. Rate of responding is most conveniently recorded in a cumulative curve. A pen moves across a paper tape, stepping a short uniform distance with each response. Appropriate paper speeds and unit steps are chosen so that the rates to be studied give convenient slopes.'

The most important of Skinner's behavioural concepts is that of **reinforcement**. This refers to outcomes of behaviour which have a strengthening effect on the behaviour. A hungry rat who randomly presses a lever and receives a food pellet is more likely to press the lever again with intention. However, Skinner is a psychologist chiefly interested in human behaviour and he believes that concepts discovered by studying lower animals can be tested at the human level. Because there are similarities among living organisms, he sees it as likely that behavioural processes found at one level may well apply at other levels. Some of his ideas have led to

the development of programmed learning, a technique which is used in some ABAL units. In a programme, the subject matter is broken down into small units. The learner should hopefully progress steadily through the material giving correct answers at the completion of each step. It is believed that each effective step in learning the material provided is reinforcing. Skinner is viewed as a controversial figure because his ideas oppose such concepts as free choice and advocate control of human behaviour by controlling the environment. To explore this further, you must read Skinner for yourself. Suggested references are given at the end of this section.

Harry Harlow was another American psychologist who studied animals in order to explore human behaviour patterns. Like Skinner he worked with captive animals in a laboratory setting and believed that apparently complex behaviours can be analysed and their development understood. He is particularly well known for researches into the mother–child relationship.

'The first love of the human infant is for his mother. The tender intimacy of this attachment is such that it is sometimes regarded as a sacred or mystical force, an instinct incapable of analysis. No doubt such compunctions, along with the obvious obstacles in the way of objective study, have hampered experimental observation of the bonds between child and mother.

Baby monkeys are far better coordinated at birth than human infants. Their responses can be observed and evaluated with confidence at an age of 10 days or even earlier. Though they mature much more rapidly than their human contemporaries, infants of both species follow much the same general pattern of development.'

'...observations suggested the series of experiments in which we have sought to compare the importance of nursing and all associated activities with that of simple bodily contact in engendering the infant monkey's attachment to its mother. For this purpose we contrived two surrogate mother monkeys. One is a bare welded-wire cylindrical form surmounted by a wooden head with a crude face. In the other the welded wire is cushioned by a sheathing of terry cloth. We placed eight newborn monkeys in individual cages, each with equal access to a cloth and a wire mother (figure 135). Four of the infants received their milk from one mother and four from the other, the milk being furnished in each case by a nursing bottle, with its nipple protruding from the mother's 'breast'.'

135   Cloth and wire mother-surrogates

These experiments showed that the monkeys took milk from both 'mothers' but that as they developed, they spent increasing time clinging to the cloth mother. In frightening situations they fled to the cloth mother for reassurance before exploring the new circumstances. Monkeys reared without their mothers or the cloth substitutes later seemed unable to form normal relationships with monkeys of either sex. Harlow concluded that 'contact comfort' was the prime requisite in the formation of an infant's love for its mother. A *Scientific American* article on this topic concludes with the words

'Finally, with such techniques established, there appears to be no reason why we cannot at some future time investigate the fundamental neuro-physiological and biochemical variables underlying affection and love.'

Konrad Lorenz wrote in his book *King Solomon's Ring*,

'Of course, one can keep animals in cages fit for the drawing room, but one can only get to know the higher and mentally active animals by letting them move about freely. How sad and mentally stunted is a caged monkey or parrot, and how incredibly alert, amusing and interesting is the same animal in complete freedom. Though one must be prepared for the damage and annoyance which is the price one has to pay for such house-mates, one obtains a mentally healthy subject for one's observations and experiments. This is the reason why the keeping of higher animals in a state of unrestricted freedom has always been my speciality.'

Discuss with your friends the advantages and disadvantages of the kind of approach used by Lorenz, Tinbergen and von Frisch compared with that of Pavlov, Skinner and Harlow. A wider reading from the references suggested on page 109 would give you a better background for discussion.

5.3.4 Observing and recording behaviour

Having considered some of the classic studies of behaviour, it is good that you should experience for yourselves a simple behavioural investigation. The first aim when studying the range of behaviour of an animal is to record it in all its detail. This can then be related to the stimuli which call forth the individual responses and then may be used to analyse separate units of this behaviour in anatomical or physiological terms. Such a detailed and complete record is termed an **ethogram**, but such complete catalogues are very rare. Nevertheless, fairly detailed records can be made by means of simple, accurate and impartial observation, meticulously recorded. It is not wise to describe a behaviour pattern after a short period of observation but much useful information can still be gained.

Inquiry and investigation opens with an investigation into the way a small mammal reacts to changes in its surroundings (practical A, page 2). This same equipment can now be used for a more sophisticated experiment which will intoduce you to techniques of observing and quantifying a piece of rodent behaviour.

Practical M: Exploration of a strange environment by rodents (open field box)

The 'strange environment' consists of an open field box divided into 25 equal squares. If a number of rodents are tested between those carrying out the practical, variations between individuals or species can be studied.

Materials

Open field box, small mammal, portable tape-recorder with bleep tape, data sheets, cling film, paper towels, dilute acetic acid

Procedure

Before starting the main investigation you need to practice the recording techniques. Measurement of mammalian behaviour is not easy because many of the categories tend to grade into one another and arbitrary distinctions must be made. You must be clear what you are looking for before beginning the investigation.

You should be able to recognise the following categories of behaviour.

Sleeping (S)	Curled up motionless with eyes closed.
No legs moving (N)	This includes freezing, sitting, movement of the head when all legs are motionless, but excludes sleeping. This category should only be recorded if the behaviour persists for more than 5 s.
Grooming (G)	This includes all licking and scratching of the body.
Locomotion (L)	This includes walking, climbing, etc.
Exploration (E)	Animal extends neck, moves its nostrils and vibrissae, may close its eyes, or rears up on its hindlegs.

If you wish to add any other category of behaviour to your data sheet, decide on it beforehand and define it.

It may be possible to add a general comment at the end of the recording activity, such as 'interest in own faeces'.

(a) Work in pairs. One person will mark the open field data sheet, the other will score for S, N, G, L and E.

(b) Place one animal in the centre square. Replace lid, if necessary.

(c) Switch on portable tape recorder which will 'bleep' every 2 s.

(d) At each bleep, note the position of the animal in the open field box by a mark in the appropriate square of the data sheet and a tick in the relevant column of the SNGLE sheet. Continue this for 5 min (indicated on tape). During 5 min of concentrated recording you should collect 150 observations on each sheet.

(e) At the end of each investigation, remove the animal gently and return it to its cage.

(f) Clean out the open field box by first removing any urine or faeces with a paper towel and then cleaning the whole area with dilute acetic acid to remove odour. Make sure area is dry before carrying out further investigations. Alternatively the base could be covered by cling film which can be disposed of after each test.

Discussion and organisation of results

The simple methods of recording and quantifying the exploratory behaviour of rodents have left you with a mass of data.

You are required to organise and analyse the data in order to present your results effectively in a visual and/or statistical form. Discuss with your partner how this may usefully be done.

1 Is movement in the open field random or non-random?

2 Can the answer to **1** be related to the properties of certain squares?

3 How varied was the behavioural repertoire of the animal?

4 Was there any evidence of fear or curiosity? Why do you say this?

5 Were there any additional significant observations?

6 Look for individual and interspecific differences and summarise these.

Show this work to your tutor.

5.3.5 Extension: Physiological determinants of animal behaviour

Laboratory investigation of animal behaviour does not stop at investigating captive animals in carefully controlled environments. The results of many behavioural observations have implied that within the brain there are genetically coded centres of motivation which control certain behaviour patterns. From the 1930s onwards, research has been carried out which has involved direct stimulation of areas of the brain. At first, this research used electrical techniques and then direct chemical stimulation.

Read *Scientific American* offprint No.464 (March 1962), *Electrically-controlled behaviour* by Erich von Holst and Ursula von St.Paul, and No.485 (June 1964), *Chemical stimulation of the brain* by Alan E. Fisher. (Both these offprints can be found in *Psychobiology: The biological bases of behaviour* by J.L. McGough, N.M. Weinbergen and R.E. Whalen.)

5.3.6 Summary assignment 7

List the scientists discussed in section 5.3 and briefly indicate their method of experimental approach and the work for which they are best known.

5.4 The classification of behaviour

There is really no place for rigid classifications when attempting to understand or describe behaviour. All behavioural acts are the result of interaction between an organism's genetic constitution and its environment. If one studies the activities of a domestic chick from hatching until it is two or three weeks old, it is possible to observe that some actions are performed well or almost perfectly the first time they occur and that others develop gradually and show improvement with practice. Sometimes the term **instinctive** or **innate behaviour** is used to describe the first type of activity, while **learning**, defined as a lasting change in the behaviour of an organism, is clearly occurring in the second kind. Closer examination of an innate behaviour in answer to the question 'How does this behaviour come about?' may blur the categories. Newly hatched chicks will peck at specks on the ground, but it has been discovered that the development of the pecking movements occurs in the embryo within the shell. Between days 3 and 12 after fertilisation the components of the pecking response gradually arise in response to tactile stimulation from movements of the yolk sac. During this time also, nervous control of the muscles is being established. After hatching, of course, the stimulus is a visual rather than tactile one, but the chick may be said to have 'practised' the motor movements before hatching. Thus, such behaviour is not really independent of experience or environment as originally thought, though clearly it is also the outcome of the inherited properties of nerve and muscle systems. The accuracy of pecking also improves with practice in the days after hatching.

Another term used to describe such behaviour is **stereotyped behaviour** or **fixed action patterns**. This behaviour results in a typical pattern of motor actions shown by all members of a given species. When the **inner readiness** to act coincides with the appropriate stimulus, then the fixed action pattern will occur.

When considering what is the stimulus which initiates such behaviour, it is often found that some factors are more important than others. These are known as **sign stimuli** or **releasers**. Sign stimuli are often identified by means of experiments using models.

A robin shows aggressive behaviour to another robin in its territory. (A territory is a defined area usually occupied by a male bird at the beginning of the breeding season and defended against incursion by other males of the same species.)

Three 'stimuli' have been placed in the territory of wild robins. These were a stuffed adult robin, a stuffed juvenile robin (with no red breast) and a bunch of red feathers as shown in figure 136. It was found that the resident males would threaten the bunch of red feathers almost as readily as they would the stuffed adult and much more readily than the stuffed juvenile.

136   Threat behaviour towards a stuffed model

SAQ 117 What feature of the stimulus appears to be most important in eliciting aggressive behaviour?

To return to the case of herring gull chicks described by Tinbergen in section 5.3.2: re-read the account of the experiments and then answer the following SAQ.

SAQ 118 What are the sign stimuli that release pecking behaviour in the chicks? What internal conditions in the chicks might also be supposed to be necessary?

The adult herring gulls are stimulated to regurgitate food when their offspring peck at their beak. The peck of the chick is the releasing stimulus for this behaviour.

Behaviour of the fixed action pattern type may appear in inappropriate circumstances. A dog or cat, for example, will turn around several times before

lying down on the floor although there is no grass to be flattened.

Geese build their nests on the ground. If an egg falls out of the nest, the sight of it will stimulate the parent to hook it under its beak and roll it back to the nest. If the egg rolls away, the bird still continues the movement back to the nest. When it, subsequently, sees the egg still outside the nest, the behaviour pattern is repeated.

Other ground-nesting birds show similar behaviour. Experimental work has shown that the larger the egg, the stronger the stimulus it provides. Figure 137 shows a gull faced with one of its own eggs and a large model egg. The bird is struggling to roll the model back into her nest while the real egg is ignored. It is often found that animals respond more readily to such **supernormal** stimuli.

137   Egg retrieval in a gull

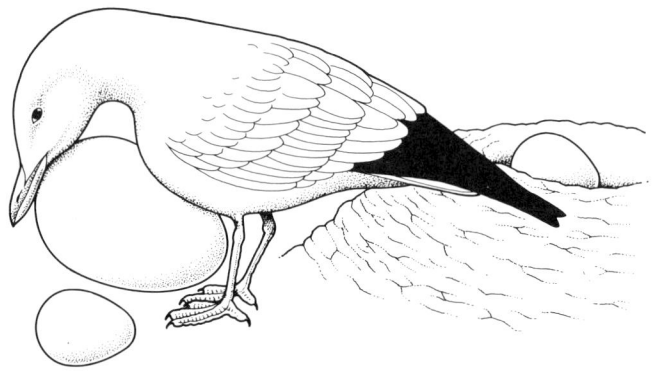

SAQ 119 What features of egg retrieval show that it is not learned?

SAQ 120 Suggest a hypothesis for why a willow warbler, a small bird, may feed a cuckoo chick in preference to its own offspring.

Behaviour is dependent on the functioning of the nervous system and its associated receptors and effectors. In the unit *Response to the environment*, the evolutionary development of nervous systems was referred to (figure 138) and is repeated from this unit.

138 Some invertebrate nervous systems

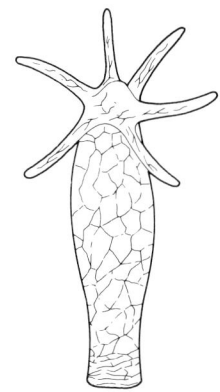

(a) Coelenterate

Hydra possesses a simple nerve net. This consists of nerve cells that may be unipolar, bipolar or multipolar. They conduct impulses slowly. It is an indiscriminate communications network in that it does not possess one-way transmission. There is also some loss of 'strength' as the impulse travels. Nevertheless, all parts of the animal are in slow communication through the maze-like web. Other coelenterates possess specialised nerve nets which also allow a speedier through conduction of impulses.

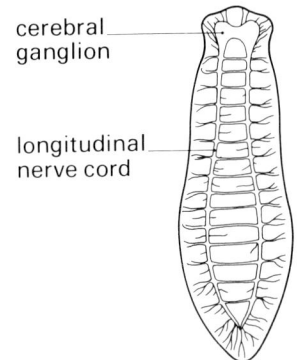

cerebral ganglion

longitudinal nerve cord

(b) Platyhelminthes

In a flatworm such as *Dugesia*, some centralisation of nervous organisation has occurred. Concentration of sensory structures on the head has led to an equivalent concentration of nerve cells in this region to form a cerebral ganglion which probably receives and integrates sensory information, not only from the head but from other regions as well, and directs and controls the behaviour of effectors.

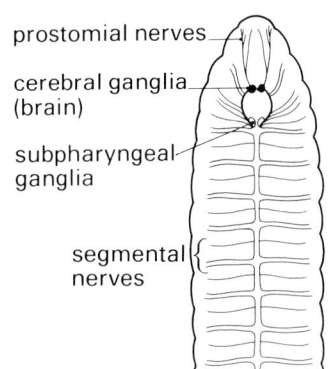

prostomial nerves

cerebral ganglia (brain)

subpharyngeal ganglia

segmental nerves

(c) Annelid

In annelids like the earthworm there is a clear-cut division into central and peripheral components of the nervous system. The cerebral ganglia may form a simple 'brain' and there is a ventral nerve cord consisting of paired segmental ganglia joined by paired connectives. Nerves containing sensory and motor fibres arise from the brain and ganglia, connect with sensory and motor systems and also connect with peripheral nerve plexuses found under the epidermis.

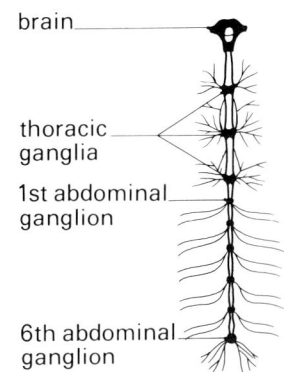

brain

thoracic ganglia

1st abdominal ganglion

6th abdominal ganglion

(d) Arthropod

In an insect, a complex brain has developed that can organise the large amount of information received through the sense organs and direct all the movements of appendages. Very complex behaviour is possible.

As the nervous system grows more complex, it makes possible a great increase in the refinement and variability of behaviour, from the simplest of stereotyped acts to the intricate and flexible responses shown in organisms with the ability to solve problems and reason. Figure 139 indicates the evolutionary changes that occur through the animal kingdom in the major types of adaptive behaviour.

SAQ 121 From figure 139, which animal group is likely to have its behavioural repertoire described as
(*a*) 'poor learners dominated by largely unmodifiable instincts, and show taxes quite clearly'?
(*b*) 'the dominant modes of adaptation are reasoning and learning and there is very little in the way of instinct or even reflex that is not greatly modified by experience; taxes are essentially non-existent'?
(*c*) 'instinctive patterns relatively simple and poorly developed and the organisms are dominated by taxes and reflexes'?

139 Behavioural trends in animal evolution

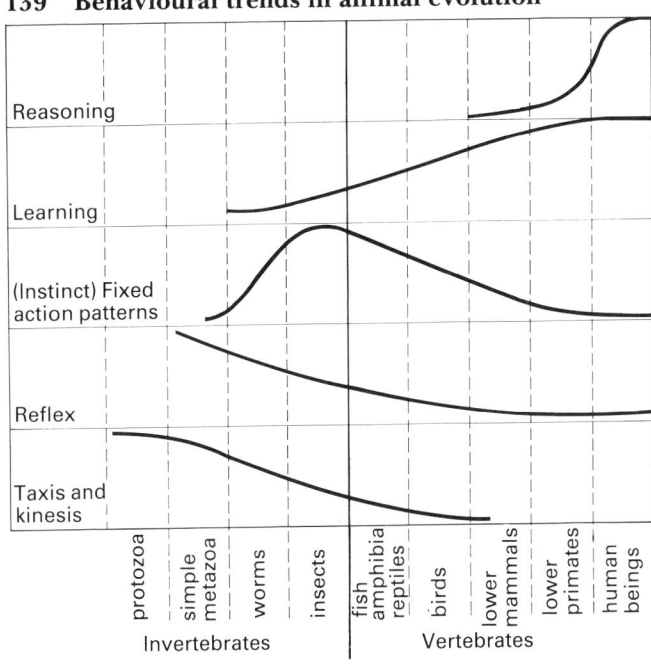

The remainder of this section investigates further these modes of behaviour.

5.4.1 Taxes and kineses

One of the simplest forms of behaviour is the orientation of an organism to some aspect of its environment. This involves locomotion in response to a specific stimulus. When the stimulus does not control the *direction* of movement but only affects the speed of movement or the frequency of turning, the response is called a **kinesis**. In this kind of behaviour, the direction of movement at any moment is completely random. *Paramecium* in a culture may be found aggregated in conditions of low carbon dioxide concentration, or moderate rather than hot or cold water. When random movements bring them into an area of high carbon dioxide concentration or hot water, they respond by alteration of their ciliary beat causing them to stop, back, alter direction and then move forward again. Figure 140 shows this so-called 'avoiding reaction' which is the basis of the behaviour. In this manner, by trial and error, the individuals of a culture of *Paramecium* tend to collect in their optimum environment.

Kineses may be of two general kinds. When the *speed of movement* is altered in relation to the strength of the stimulus, the reaction is called **orthokinesis**. When

140   Avoiding reaction in *Paramecium*

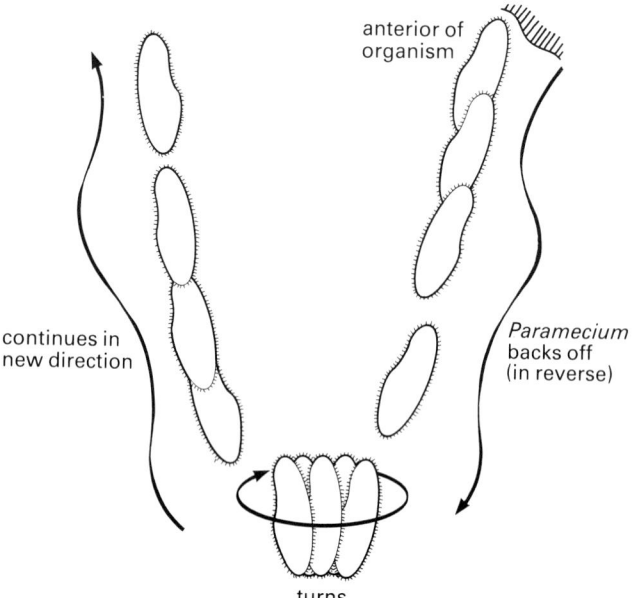

the *rate of turning* is altered in response to a stimulus it is described as **klinokinesis**. The graph in figure 141 shows the effect of light on the locomotion of the planarian *Dendrocoelum* sp. The animals were illuminated from directly overhead and all reflected light was eliminated. In another investigation where a gradient of overhead light intensity was provided and the planarians at first scattered evenly over the surface, they were found collected in the dark end after about one hour.

SAQ 122 What type of kinesis is displayed by *Dendrocoelum* in the investigation reported in figure 141?

141   Graph showing the relationship between light intensity and the rate of change of direction in *Dendrocoelum*: A–B, darkness; B–C, light on (from Ullyot, *J. exp. Biol.*, 1936)

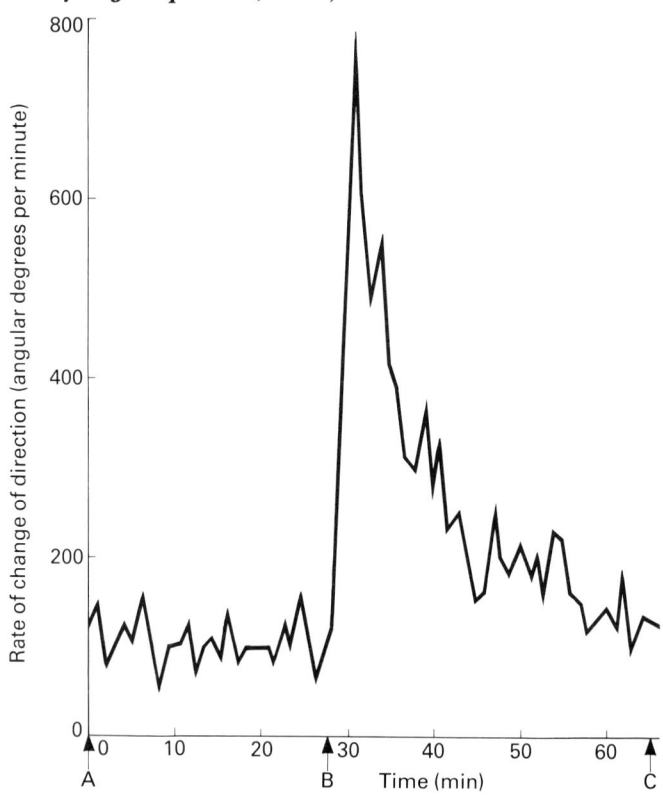

SAQ 123 Is the rate of turning related to the direction of the stimulus?

SAQ 124 Briefly describe the effect of light on the behaviour of the planarian.

SAQ 125 Suggest a reason for the gradual decrease in turning activity.

Woodlice are generally found in damp places under stones and decaying wood and leaves. Choice chamber investigations offering wet and dry environments reveal that animals in the moist chamber soon come to rest and move about little. (You may have carried out such an investigation when studying the unit *Inquiry and investigation*.) In the dry chamber the woodlice are constantly moving. Figure 142 shows the results of investigations into the locomotory activity of a woodlouse, *Porcellio scaber*. Each point represents the percentage of animals not moving in one experiment at constant humidity. The line joins the average points.

142   Locomotory activity of the woodlouse *Porcellio scaber* in relation to humidity (from D.L. Gunn, *J. exp. Biol.,* 1937)

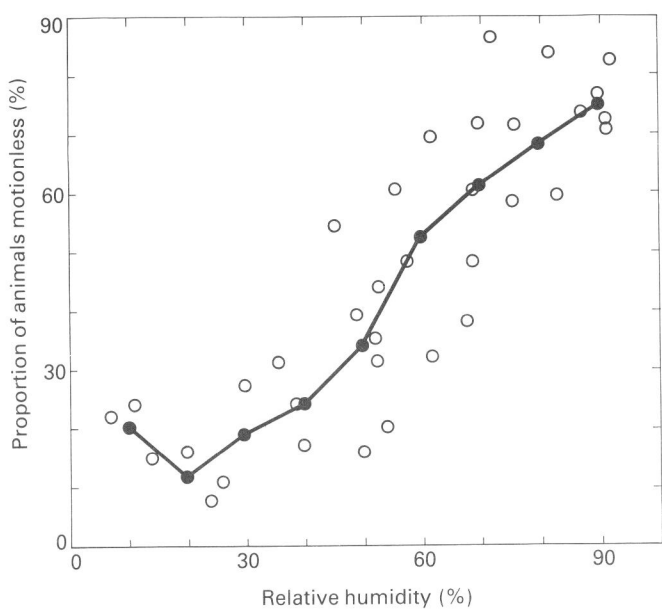

SAQ 126 Use this data to suggest a hypothesis to answer the question 'Why are woodlice usually found in damp places?'

SAQ 127 What evidence is there that a kinesis is involved? Which type of kinesis do you suspect?

SAQ 128 What is the survival value of this behaviour?

In the unit *Response to the environment* positive **phototaxis** in *Euglena* was described. The organism moved directly towards the light source. In a taxis, the direction of movement is related to the direction of the stimulus. Taxes are named after the nature of the stimulus and described as positive or negative; thus chemotaxis, geotaxis and phototaxis.

There are two different ways in which an organism may demonstrate phototaxis, for example, when a single light provides the stimulus, the animal may move straight forward with its body axis in line with the light rays. It orientates directly by *simultaneous* comparison by means of bilateral light receptors. This is a **tropotaxis**. In other organisms regular symmetrical deviations occur from the direct path. Such organisms do not possess bilateral photoreceptors and response to the stimulus involves comparison of light intensities at *successive* points in time. This method of response is a **klinotaxis**: it involves a turning of the light-sensitive area.

SAQ 129 Which type of taxis is likely to be displayed by *Euglena* when orientating towards light?

5.4.2 Investigation of some invertebrate responses to light

This section consists of a circus of four practicals. Do as many of them as you can. If you are in a group, they may be shared out among you. For each practical, carry out the procedure as indicated. Check that you are eliminating as many variables as possible. Criticise the methods used and suggest your own improvements.

Be prepared to present the aims, method, results and discussion of any practical you have done to the rest of your group. Your tutor will advise you which practicals to prepare for presentation. For these practicals the room should be darkened.

Practical N: Orientation in flatworms

Materials

Petri dish, graph paper (2 A4 sheets), paintbrush for transferring flatworms, dish of 10 flatworms in pond water (previously kept in the dark), scissors, sellotape, lamp with 60 W bulb which can be suspended above the petri dish at varying distances, pie-dish (20 × 25 cm) half-filled with pond water, piece of opaque cardboard to cover half pie-dish, stopwatch or clock

Procedure

Part A

(a) Place 10 flatworms in a pie-dish of pond water and ensure they are evenly distributed.

(b) Cover half the pie-dish with opaque card and shine a lamp on the dish from a height of 20 cm.

(c) Observe the animals and record what happens.

Part B

(a) Cut five circles of graph paper equivalent in size to the petri dish.

(b) Stick one circle of graph paper to the base of the petri dish. Place another at its side. (The remaining three will be used later in points (f) and (g).)

(c) Half-fill the petri dish with pond water.

(d) Transfer one flatworm to the petri dish and leave for five minutes in semi-darkness to acclimatise.

(e) After five minutes, record the movements of the flatworm by drawing a copy of its tracks on the graph paper circle at the side of the petri dish. Indicate 30 s intervals on the track. Continue for 3–4 min.

(f) Place a lamp 60 cm above the petri dish and turn it on. Repeat the activities outlined in point (e).

(g) Repeat this procedure with the lamp at 30 cm and 15 cm distances from the petri dish.

(h) Now repeat the whole procedure from points (d)–(g) for five other flatworms. This may be done by correlating the work of other groups.

(i) Devise a way to compare quantitatively the degree of turning of the flatworms in different conditions.

(j) Draw histograms to show (i) the degree of turning and (ii) the speed of movement in relation to light. Strength of light should be recorded along the x axis.

Discussion of results

1 In part A of the investigation, in which conditions did most of the flatworms eventually congregate?

2 From your investigations in part B, did light intensity appear to affect either of the following: (a) distance moved in five minutes (i.e. speed of movement), (b) degree of turning?

3 Do you have any evidence that the response of the flatworms to light was either kinetic or tactic? What further evidence would you need to confirm your hypothesis?

Show this work to your tutor.

Practical O: Response to light by fly larvae

Warning: be careful to handle these animals as little as possible. Even slight damage will affect your results. Check also that desiccation does not occur as this also affects behaviour.

Materials

2 large sheets of paper (double A3), methylene blue dye, 2 bench lamps or torches, fly larvae (maggots), paintbrush for handling animals, binocular microscope

Procedure

(a) Arrange the lamp to give a single, directional beam across the paper. Mark the outline of the beam on the paper.

(b) Place a maggot dipped in methylene blue in the beam of the lamp, fairly close to its source.

(c) Observe its movements and behaviour. Make a record of these observations.

(d) Repeat with at least two other maggots.

(e) Take a second sheet of paper and set up the lamp as in (a).

(f) Arrange a second lamp to produce a beam at right-angles to the first. Switch it off.

(*g*) Place a maggot as in (*b*).

(*h*) When the maggot has travelled some distance, switch off the first lamp and switch on the second lamp.

(*i*) Transfer the record of the maggots' paths to a smaller sheet of paper for storage in your file or practical book.

Observe the larvae under a binocular microscope. Note any structures you see which might be light receptors.

Discussion of results

1 Is the behaviour of the blowfly larvae in relation to light a tactic or kinetic response? Attempt to classify the response. Explain your answers.

2 Where is the photosensitive region in these maggots? Did you detect light receptors?

3 Suggest a hypothesis for the reception of the light stimulus in these maggots. Explain the reasoning that led you to make this suggestion.

Show this work to your tutor.

Practical P: Response to light by woodlice and/ or larvae of *Tenebrio molitor* (meal worm)

Materials

Large sheets of paper (double A3), 2 bench lamps or torches, soft pencil, woodlice or *Tenebrio* larvae, paintbrush, hand lens or binocular microscope, non-toxic enamel paint

Procedure

(*a*) Arrange your light source to give a single directional beam across the paper. Mark the outline of the beam on the paper. Note the warning given in practical O.

(*b*) Place a woodlouse (or *Tenebrio* larva) in the centre of the paper (a pooter may be used) and trace its resultant path with a soft pencil. Repeat at least five times.

(*c*) Take a second sheet of paper and set up the lamp as in (*a*).

(*d*) Arrange a second lamp to produce a beam at right-angles to the first. Switch it off.

(*e*) Start a new series of trials, but when the animal has moved some distance, switch off the first light and switch on the second light.

(*f*) Identify the eye of a woodlouse using figure 143.

(*g*) Paint over *one* eye and repeat (*b*).

143   A woodlouse, showing eyes

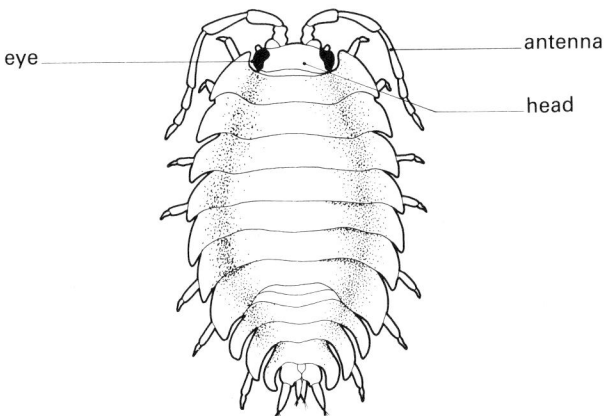

Discussion of results

1 What kind of response to light is shown by the organism?

2 What type of light receptors are present?

3 Suggest why the animal moves forwards (in terms of receptors and possible role of CNS).

4 Why does the animal turn (as above)?

5 Explain the survival value of this behaviour in relation to the natural habitat of the organism.

Show this work to your tutor.

Practical Q: Orientation in brine shrimps
(*Artemia* sp.)

Materials

Small tank or beaker of sea water with brine shrimps, facilities for blackout or use of a dark room, powerful torch or bench lamp, tripod, binocular microscope or hand lens

Procedure

(*a*) Observe the orientation of the dorsal surface of the body of the shrimps in daylight, that is with light directed from above. (Avoid doing this investigation near a window.)

(*b*) Transfer the culture of shrimps to a tripod in a dark room. Shine the light on the culture from below the beaker. Observe the orientation of the dorsal surface of the animals again.

(*c*) If you have access to a dark room with infra-red light, investigate the orientation of the dorsal surface of the shrimps under these conditions.

(*d*) What photoreceptors are possessed by *Artemia*? Use the hand lens to investigate.

(*e*) Record all your observations.

Discussion of results

1 Does light affect the behaviour of the brine shrimps?

2 Is the direction of movement of brine shrimps related to the direction of the light source?

3 What categories of behaviour are displayed by the shrimps in this investigation?

4 Suggest ways in which the behaviour patterns you have observed may be advantageous to the survival of the brine shrimps.

Show this work to your tutor.

5.4.3 Reflexes

Reflex actions were studied in section 4 of *Response to the environment*. They are the simplest unlearnt responses found in organisms possessing a nervous system. Their nature is determined by an inherited pattern of receptors, nerves and effectors. Reflexes may involve various levels of the nervous system and vary in complexity. The knee-jerk reflex is thought to be organised within a few sections of the spinal cord. When flexion and extension reflexes are coordinated to produce walking, a wider area of the spinal cord is involved and also the midbrain.

The concept of the reflex is important in the description and analysis of behaviour. At one time, it was believed that all behaviour might eventually be understood in terms of a complex series of reflexes. Even learning was seen as a combination of innate and conditioned reflexes, but this view is not generally accepted today.

5.4.4 Conditioned reflexes: classical conditioning

In section 5.3.3 you read of Pavlov's well-known investigation into conditioned reflexes in dogs. A process in which a stimulus that was previously unrelated to a particular response eventually produces that same response as a result of training is called **conditioning**. This is a passive form of learning since the conditioned stimulus is imposed by the experimenter or environment.

In Pavlov's experiment the metronome sound is known as the conditioned stimulus and salivating in response to its clicks is called the **conditioned reflex**.

5.4.5 Operant or instrumental conditioning

Some types of behaviour which apparently occur spontaneously may affect an animal's situation, for instance by bringing rewards or avoiding punishment. This reinforces the behaviour and so the animal does it more often. This is called **operant** or **instrumental conditioning**. This kind of trial-and-error process is an important element in learning. Section 5.3.3 introduced you to the work of Skinner with pigeons and rats in a Skinner box.

SAQ 130 What is the important difference between classical and operant conditioning?

5.4.6 Habituation

A loud noise or flash of light causes a kitten to run away. If the kitten is subject to repeated noises or flashes, the response becomes less and less. Repeated stimulations associated with diminution of response is described as **habituation**.

This is perhaps the simplest kind of learning and, in some ways, represents the ignoring of stimuli that are of no 'significance' in the life of an animal. In contrast, most other learning is concerned with the strengthening of significant responses so that they may be more readily produced.

5.4.7 Trial-and-error learning

This is a form of instrumental conditioning in which an animal is given a choice of stimuli to respond to. After several trials, it almost always makes the choice associated with reward or avoids a choice associated with punishment.

This type of behaviour can be illustrated when a hungry rat is placed in a maze with food at the end-point (figure 144). The animal is allowed to run through it many times. With successive trials, the speed and accuracy with which it reaches the end-point increases.

5.4.8 Latent learning

Rats can be trained to run through a maze with a minimum of wrong turnings (errors) using a reward of food at the end-point.

In an investigation into the effects of food rewards on learning, three separate groups of rats were tested in a simple maze over a period of 15 days.
Group 1 rats received a food reward when they reached the end of the maze.
Group 2 rats received no food reward at any time.
Group 3 rats received no reward for the first eight days, but from the ninth days onwards were rewarded with food at the end of the maze.

The results are shown in figure 145.

144   A maze

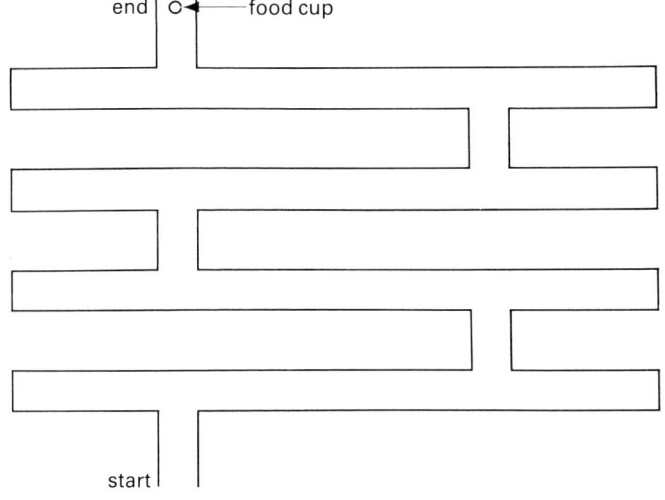

145   Maze-learning in rats

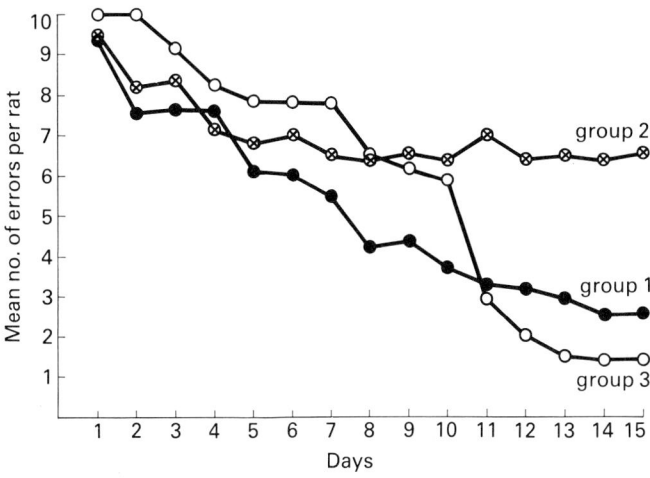

SAQ 131 How long was the training period for each group?

SAQ 132 Suggest a reason for the improvement shown by all three groups during the first three days of the investigation.

SAQ 133 Briefly describe the difference in the results of groups 1 and 2.

SAQ 134 Propose an explanation for the rapid improvement of the results of group 3 rats after day nine.

Many animals explore their surroundings and store the information in their memory. Later, this may have useful consequences, such as they may be able to escape from, or hide from, predators more easily.

This making of associations without immediate reward or punishment is known as **latent learning**.

5.4.9 Imprinting

When a young bird or mammal hatches or is born, it will follow around the first large moving object that it sees. This is usually its mother. This behaviour is known as **imprinting**.

SAQ 135 In what way could imprinting be important in terms of survival?

Section 5.3 contains an account of Lorenz's observations on Muscovy ducks and Greylag geese. Reread the passage now.

SAQ 136 At what stage in an animal's life does imprinting occur? Does an animal imprint upon more than one object?

SAQ 137 There is also evidence that imprinting determines subsequent behaviour. What was unusual about the behaviour of the male Muscovy duck when it reached maturity?

SAQ 138 Explain what Lorenz meant when he said 'The process of acquiring the object of a reaction is, in many cases, completed long before the reaction itself has become established...'

Later investigations by other behavioural scientists made clear further details of this process. Figure 146 shows the average test score of following responses to a model made by young mallard ducklings.

SAQ 139 What appears to be the optimum age for imprinting?

146 Critical age at which ducklings are most strongly imprinted (data from Hess)

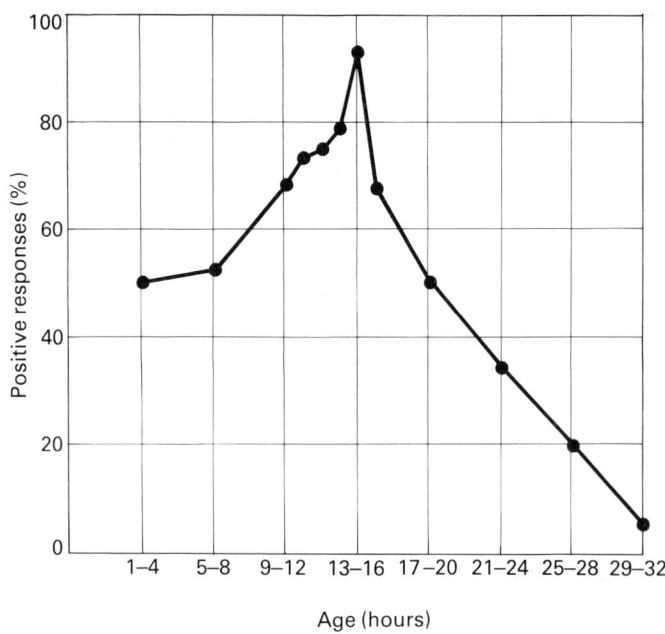

147 Strength of imprinting related to distance travelled (data from Hess)

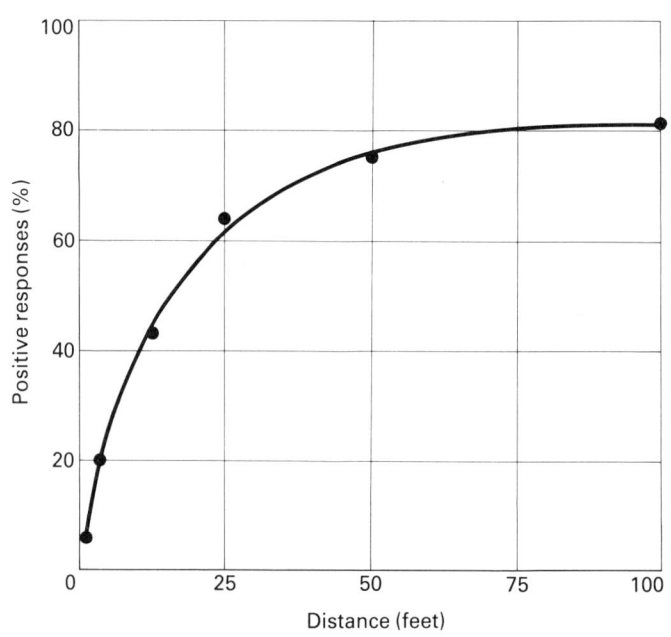

Figure 147 shows the strength of imprinting related to the distance actually travelled by a duckling while following the object on which it is becoming imprinted. For imprinting to occur, the 'object' must be moving. The positive responses of the y axis relate to later positive responses after the critical period for imprinting.

SAQ 140 (*a*) What do these results reveal?
(*b*) What factor does this suggest is also involved in the process of imprinting?

5.4.10 Insight learning or reasoning

The ability to solve a new problem using more than trial and error or habit is known as **insight learning**. The solution often appears to come to the animal suddenly.

If chimpanzees are presented with a bunch of bananas too high to reach, they will pile up boxes to stand on or fit two sticks together to pull them down. If this is done without previous experience and without trial and error, it is an example of insight learning. The chimpanzees would, of course, have benefited by previous experience of playing with boxes and sticks (latent learning) and trial and error is used when building a stable pile of boxes. However, they are able to use knowledge obtained in one context and apply it in another.

One classic problem used to investigate problem-solving in animals is the detour problem illustrated in figure 148.

148 Detour problem in which an animal must first go away from the food in order to reach it

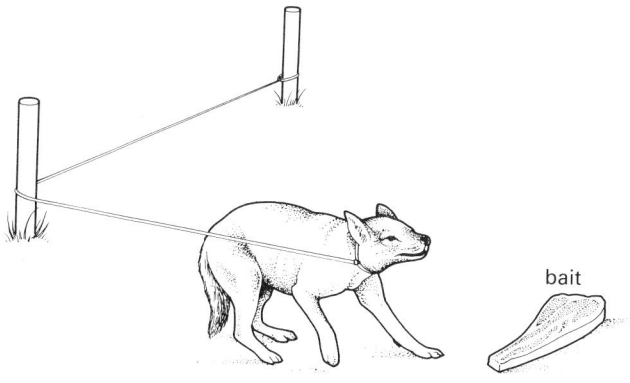

bait

Of the variety of animals tested, only monkeys and chimpanzees show much success when first faced with the problem, that is show insight learning. Rats, dogs and racoons learn rapidly, but through trial and error. Fish and birds learn more slowly.

Human beings generally use insight to solve problems. One form of this is the 'brain-wave' or suddenly 'getting it' (also described as the 'aha!' reaction). Concept formation is another aspect of insight learning. Harlow's work with monkeys and children suggests that the ability to solve complex problems is not inborn but is gained by practice in problem-solving. He does not consider trial and error and insight as separate, but different phases of one long continuous process. They represent the orderly development of a learning and thinking process.

Figure 149 shows a young Rhesus monkey taking part in an oddity test. By presenting the monkey

149   Monkey solving oddity problem

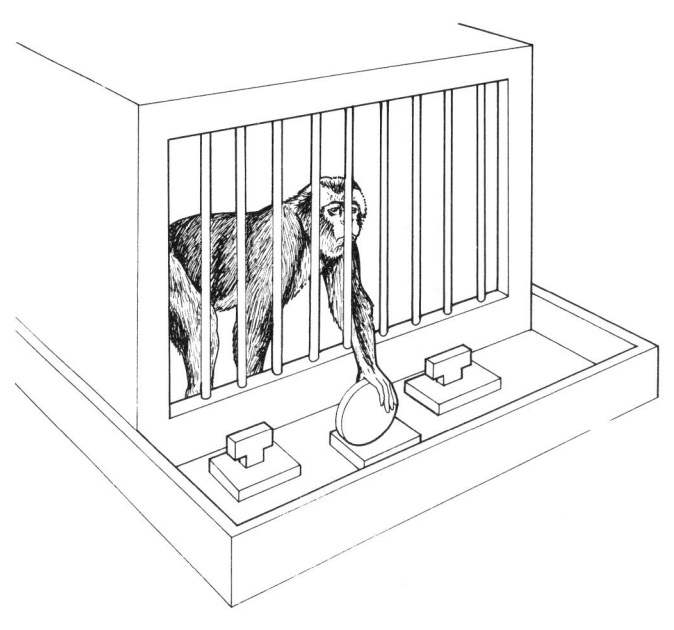

with a choice between three objects, two similar and one different, the animal gradually learns that food will always be found under the 'different' object. Here, the animal learns that the shape is not important but its relation to the other two objects. At first, trial and error is clearly involved, but eventually the correct choice is made first nearly all the time. Harlow referred to a concept such as oddity as a 'learning set'. He sees that an individual may organise simple learning sets into more complex patterns which can then be transferred as units to new situations.

5.4.11 Practical investigations of learning behaviour

You may already have ideas for behavioural investigations you can design for yourselves. For example, dog-owners may wish to try the detour problem. Most maze-running investigations involve depriving an animal of food for about 48 h and simpler mazes used with worms or planarians may involve administering mild electric shocks. The same factors generally hold for investigations in operant conditioning. For food to be the reinforcement, an animal must be hungry and negative reinforcement takes the form of punishment. The following investigations have been selected because they do not involve organisms in any type of 'hardship'. (A licence is required if an experiment is to place vertebrates under any degree of distress such as a period of starvation.)

Practical R: Withdrawal of tentacles by snails

Materials

Petri dish covered with thin film of water, terrestrial snail (*Helix* is a good subject), glass rod, stop-watch

Procedure

(*a*) Touch the dorsal surface of the head in the midline with the end of the glass rod.

(*b*) Observe and record nature of response.

(*c*) Repeat the procedure at 1-min intervals for about 15 min. It is important to keep the stimulus strength as constant as possible.

(*d*) Record the percentage withdrawal of the snail (this must be a subjective estimation, so have two observers). Record also the time taken for the tentacles to be fully extended again. (Emergence of eyes may be a useful criterion.)

If the initial response appears to be too big, try either a weaker stimulus or 2-min intervals.

(*e*) Display your results by means of histograms. Plot separately the *extent* (%) and *duration* (s) of response against time in minutes (successive stimulations).

Discussion of results

1 What difficulties did you experience in carrying out this practical?

2 Explain what category/ies of behaviour are displayed by the snail.

3 Suggest one way in which the behaviour shown might be an adaptation to survival.

4 What factors might affect the behaviour of the snail?

5 Outline the practical procedure you would use to investigate the effects of one of these factors.

Show this work to your tutor.

Practical S: Typing in human beings

Materials

Typewriter, typewriting paper, stop-watch

Procedure

(*a*) Work in pairs.

(*b*) One member of the pair should type the sentence 'Hey diddle diddle, the cow jumped over the moon.' The partner should record the time taken.

(c) Repeat 100 times. If possible, note when any group of letters are fired off more rapidly for the first time.

(d) Draw a graph of trial number against time.

Discussion of results

1 Explain what category/ies of behaviour were illustrated in this activity.

2 Account for any irregularities in the curve.

3 Suggest five factors which might affect the speed of typing.

4 Account for individual differences between subjects when class results are compared.

Show this work to your tutor.

The unit *Inquiry and investigation* includes a practical involving human learning (see p. 47, practical J: The star drawing practical). If you have not tried this before, it would be appropriate here.

Practical T: Human learning with a pencil maze

This practical has two purposes, the first as an investigation of human learning and the second as an exercise in experimental design. (This practical could be extended beyond the laboratory session to include tests on friends and family.)

Materials

Pencil maze, pencil, sheets of A4 paper, stop-clock or watch, blindfold or other means of obstructing view of the maze

Procedure

(a) Work in pairs

(b) One member of the pair will attempt to follow the maze while the other records errors and the time taken.

(c) Decide on the number of trails to be made per learner.

(d) Present results in graphical form.

This practical could be used to investigate differences/transfer between left and right hands, changing orientation of the maze, the effects of age, sex, conditions under which the maze is 'run', 'artists or scientists', use of instrument or finger, strategies of learning, and so on.

(e) Write a concise account of your methods including the control of variables.

(f) Write a careful discussion of your findings.

Show this work to your tutor.

5.4.12 The nature of memory

A whole new dimension was added to behavioural adaptation when animals evolved the capacity to learn and to modify their responses in the light of past experience. Learning makes it possible to adjust behaviour as the environment changes and this flexibility opens new possibilities and greatly enhances the likelihood of survival. However, for an animal to learn it must be able, in some way, to record the new experience or information, that is, remember it. But what is memory? How is information stored in the nervous system of an animal and retrieved in appropriate situations?

AV 2 investigates what is currently known about memory.

AV 2: The mechanisms of memory

Materials

VCR and monitor
ABAL video sequence: *The mechanisms of memory*
Worksheets

Procedure

(*a*) Check that you have all the relevant materials for this activity.

(*b*) Check that the video cassette is set up ready to show the appropriate sequence – *The mechanisms of memory*.

(*c*) Start the video and stop it at the appropriate point to fill in the worksheets.

(*d*) If you do not understand something, stop the video, rewind and study the relevant material again before consulting with your tutor.

(*e*) If possible, work through the video and worksheets with a small group and discuss the material amongst yourselves.

5.4.13 Summary assignment 8

1 Write definitions of the following terms and provide an example of each.

fixed action pattern
sign simulus or releaser
supernormal stimulus
reflex behaviour
kinesis
taxis
conditioned reflex
operant or instrumental conditioning
habituation
imprinting
trial-and-error learning
latent learning
insight learning

2 Distinguish between the terms **innate** and **learned** behaviour. Write down two objections that are made to the use of the term 'innate'. Explain why it can still be convenient to use the description 'innate'.

Self test 6, page 114, covers sections 5.2–5.4 of this unit.

5.4.14 Extension: The evolution of intelligence

At the beginning of this section, the repertoire and flexibility of an animal's behaviour was related to the complexity of its central nervous system and associated receptors and effectors. It is clear that animals have limits to their capacity to learn and perform, and that these capacities vary between species. What is the nature and the basis of these differences?

M.E. Bitterman has attempted to study species differences by training animals to solve problems concerned with habit reversal and probability learning. He has applied his findings to ideas about the evolution of intelligence in the vertebrates.

Read *Scientific American* offprint No. 490 (January 1965) *The evolution of intelligence* by M.E. Bitterman. (This offprint is included in *Psychobiology–The biological bases of behaviour*, published by W.H. Freeman & Co.)

Answer the following questions.

1 What basic hypothesis did Bitterman set out to investigate?

2 What type of behaviour pattern is used to train the animals?

3 What type of learning is actually required of the animals?

4 Good experimental design requires the control of variables. What problems did the author meet in this respect and how did he attempt to overcome them?

5 How successful is the author in detecting an evolutionary trend?

6 Can you link the behavioural findings with any evolutionary changes in brain structure? (Refer to section 4.10.3 of the unit *Response to the environment*.)

7 Do you consider that further investigations might reveal significant differences in intelligence between monkeys and rats? Justify your answer.

Show this work to your tutor.

5.5 The internal environment

The same stimulus presented to the same animal at different times will not always give rise to the same response. You have seen that there is a 'critical period' for imprinting to occur. A number of theories are put forward to account for this, including the development of fear and increasing sociability. Another suggestion is that 'readiness for imprinting wanes on the basis of maturational factors alone.' **Maturation** is associated with the continued development of an animal's nervous system which leads to improvement in the performance of a behaviour, the appearance of a new behaviour, or the loss of sensitivity to stimulus. Practice is not involved in this phenomenon, but in the majority of cases the process of maturation and practice interact. This is well illustrated by the development of the pecking response in newly hatched domestic chicks. Figure 150 shows the results of an investigation into the accuracy of pecking at grains of millet on a black floor.

150   Chicks: average number of misses out of 25 pecks

Age in hours	No practice	12 h practice
24	6.04	—
48	4.32	1.96
72	3.00	1.76
96	1.88	0.76
120	1.00	0.16

SAQ 141 How does practice appear to be related to maturation in this example?

Learning, of course, is defined as a lasting behavioural change which may alter a response, and the effects of learning have been considered in the previous sub-section.

Motivation also affects the nature of a behavioural response. This is a difficult concept and is concerned with the factors that determine the direction of an individual's behaviour, the strength of a response. In investigating motivation, the hypothalamus of the brain has been found to play an important role and specific mechanisms have been found for thirst, sexual, emotional and maternal behaviour and sleep. These findings have led to the idea of basic **drives** – a striving to attain a goal. A hungry animal usually has to seek out food using a variety of behaviour patterns. When the food (the goal) is located, the variable searching behaviours give way to behaviours that tend to be more stereotyped, that is the actual mode of feeding or consummatory act. This is generally followed by a period when the animal is no longer responsive to stimuli relating to food: the drive has been reduced and motivation in this area is low.

The notion of drive is dependent on mechanisms for monitoring the internal state of the animal. Sensory receptors within the body of an organism detect internal temperature, concentration of carbon dioxide in the blood, pressure on the bladder, and so on.

SAQ 142 A young baby awakes and cries. Its mother feeds it and returns it to its cot. It soon goes to sleep. Explain this behaviour in relation to what has been said about motivation and sensory receptors.

Perhaps the greatest internal influence on the behaviour of animals (apart from the CNS) is the action of hormones.

5.5.1 Hormones and behaviour

Of all the internal factors that influence behaviour, the actions of hormones are the most clearly understood. It is possible to precisely measure hormone levels in the body and to relate these to behavioural changes. Even more significantly, endocrine organs can be removed with little damage to the animal. It is then possible to examine the role of a specific hormone by injecting known quantities of it into the animal.

Much research has been carried out into the effects of the sex hormones on behaviour. Testosterone, for example, is known to influence not only sexual behaviour but also aggression. Research on chicks and mice has revealed that it may increase the persistence with which individuals search for a particular kind of food. Thus, the one hormone has effects on several types of behaviour. Most hormones influence behaviour over a relatively long time scale.

SAQ 143 Name a hormone which is an exception to this last statement. Indicate the kind of behavioural activity its secretion would be expected to produce.

Hormones do not work in isolation but *interact* in complex ways with external stimuli. The secretion of sex hormones in most birds and many mammals of temperate regions is controlled by the photoperiod. Increasing day length in spring triggers the production of sex hormones. A hormone can seldom produce a behaviour on its own without the appropriate external stimulus: in the case of sexual activity, the presence of a possible mate.

A very detailed analysis of the way hormones, external stimuli and behaviour interact has been carried out studying the reproductive behaviour of the ring or Barbary dove (*Streptopelia risoria*). The work was begun by Daniel Lehrman (see Further Reading) and continued by many other workers. Figure 151 shows the principal connections between physiological changes, external stimuli and behaviour.

SAQ 144 Which hormones influence the female bird?

SAQ 145 What activity of the male bird prepares the female for nest-building?

SAQ 146 What triggers incubation?

SAQ 147 What factors are necessary for crop-milk production?

There is a marked rise in testosterone production in the male after pair formation, but many changes in the male's behaviour appear to be triggered directly by the activities of the female. A general receptivity to the stimuli seems to be the chief effect of the hormone and the stimuli coordinate his behaviour. Prolactin, stimulated by incubation, has a similar effect on both sexes.

151 Analysis of factors affecting the reproductive behaviour of female ring doves

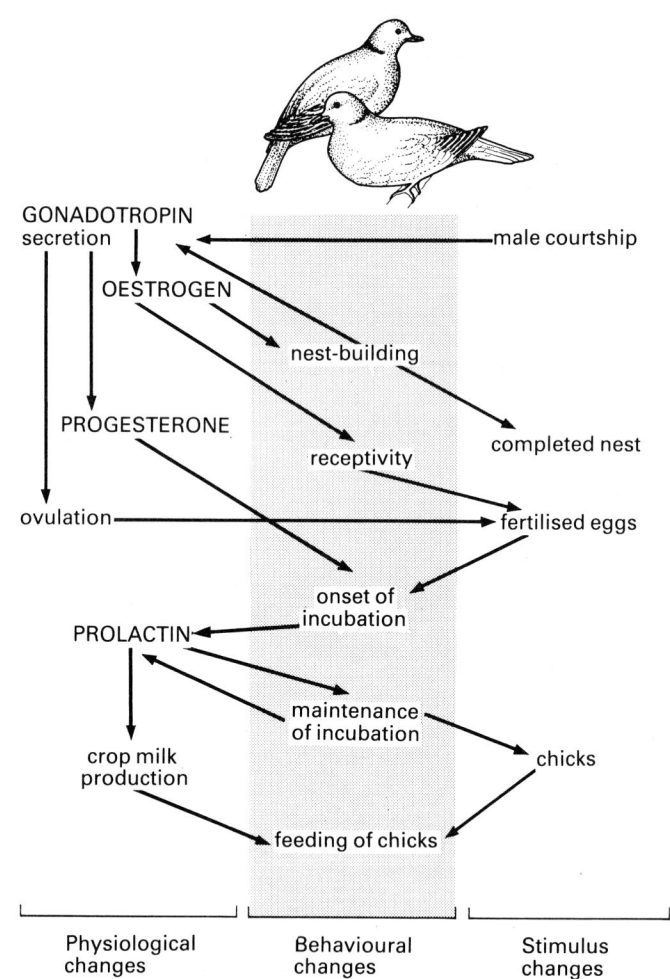

5.5.2 Biological rhythms (endogenous rhythms)

The behaviour of many animals follows a rhythmic pattern. There may be annual rhythms associated with courtship, mating, raising young, hibernation, and so on, or daily rhythms associated with periods of rest and activity. Other animals show lunar or tidal rhythms. These rhythms may simply reflect correspondence to relevant environmental factors such as food availability, but, in many cases, there is evidence that they are controlled by internal (**endogenous**) rhythms which function largely independently of environmental cues.

152   Effect of uniform conditions on rhythms of flight activity in the mosquito

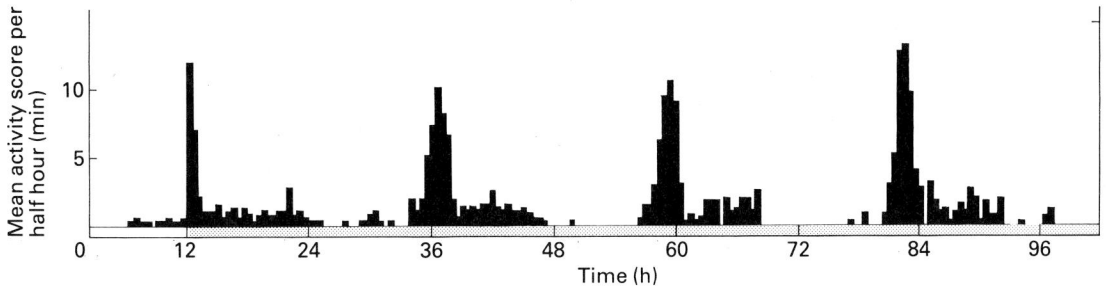

The mosquito *Anopheles gambiae* is a nocturnal species but with noticeable peaks of activity at dusk and dawn. When removed from conditions of 12 h light, followed by 12 h darkness, to conditions of uniform dark, the dusk or 'light-off' peak persists and recurs with a rhythm of close to 23 h for several days, as can be seen in figure 152. This suggests that the animal has an internal timing device or clock which enables it to regulate its own behaviour patterns. When these rhythms are allowed to run in uniform environmental conditions they are often a little over or below a 24-hour cycle. Thus, they are called **circadian rhythms** (*circa* – about, *dies* – day). In natural conditions, the environment appears to adjust the animal's biological rhythm to a more precise 24-hour cycle. Light is an important factor in this regulation.

It is not known how the internal biological clock operates, although the central nervous system must be involved in some way.

Sleep is an example of a behaviour which normally occurs rhythmically. The sleep requirements of different species varies greatly. Dolphins and shrews have never been recorded as sleeping while a sloth may sleep 20 h a day. Some physiological activities occur mostly during sleep, but some have suggested that sleep is primarily a mechanism to keep animals safe at times of day when they are ill-adapted to function. If you have ever experienced 'jet lag' you will recognise that the timing of sleep is determined both by environmental change and by an endogenous circadian rhythm which needs to be reset when you move from one time zone to another.

5.5.3 Photoperiodism

In higher latitudes where regular annual changes in the length of daylight and darkness occur, the photoperiod has a major synchronising effect on annual rhythms. The increasing day-lengths that occur in early spring stimulate many birds to become territorial and begin nest-building. These activities are related, as you have just seen, to an increase of sexual hormone levels. Figure 153 shows the rate at

153   Rate of testicular development as a function of daily photoperiod in six species of bird

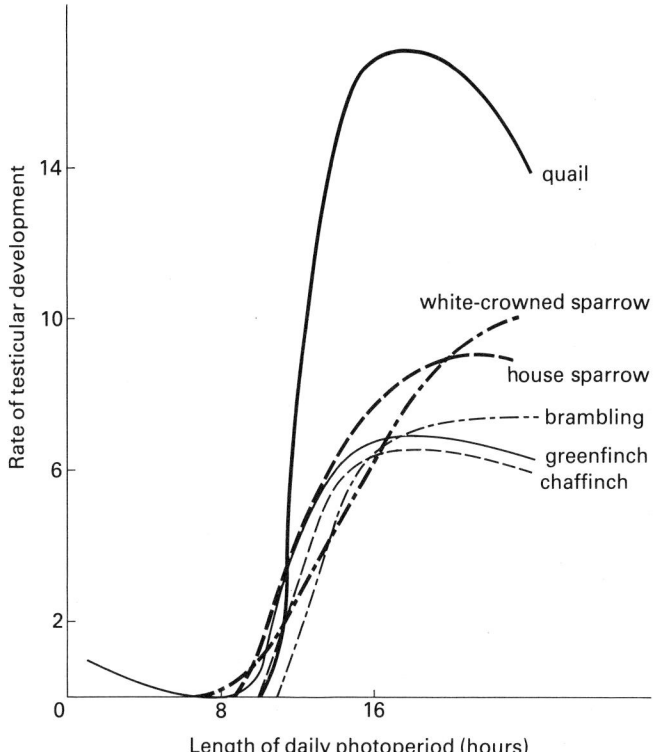

which the testes develop related to the length of the daily photoperiod under experimental conditions.

SAQ 148 It is important that animals produce their young at a favourable time of year. Relate the data shown in figure 153 to this fact.

SAQ 149 If the photoperiod is regarded as an environmental synchroniser, what aspect of the photoperiod is likely to stimulate gonad growth in sheep? (The gestation period for sheep is 150 days.)

Annual migrations of birds are also related to photoperiod and differences of a few minutes in photoperiod length have been shown to be important. Exactly how daylight is 'measured' is uncertain, but one theory relates the timing to circadian rhythms and suggests that there is a rhythm of photosensitivity so that when light is experienced at the proper phase it acts as a stimulus to the relevant process.

Recent studies on the human **pineal gland** (an organ whose function has puzzled physiologists for many years) have suggested that it has a role in circadian rhythms, particularly those regulated by light and darkness. The photoperiod has been shown to govern the endocrine output of the pineal gland in human and animal experimental subjects. Light appears to suppress the activity of the gland. In seasonally breeding animals, the pineal gland acts as an environmental link between seasonal photoperiodic changes and the neuroendocrine control of reproduction.

5.6 Communication

Communication may be defined as passing information from one animal to another by means of signals which have evolved for this purpose. Communication will influence the behaviour of the animal receiving the signal and almost all species of animals have communication systems which convey specific information and aid interaction between species members. The signals may be a structure or a chemical or a behavioural event. Communication also occurs to a lesser extent between species as the following examples show.

154   Cleaner fish at work

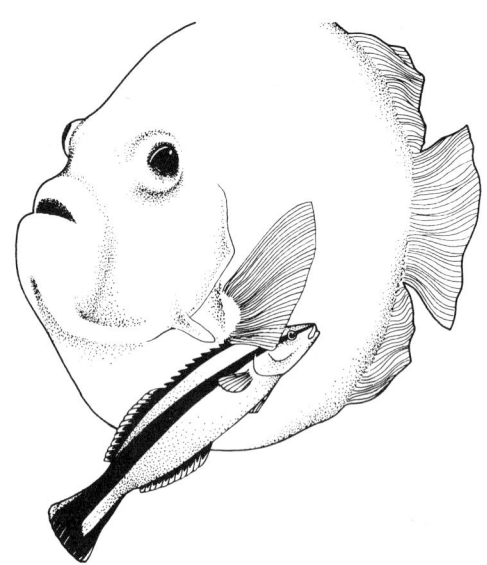

The scales of fish often become encrusted with other organisms and are cleaned by smaller 'cleaner' fish. These 'cleaners' come from unrelated families but all are characterised by black longitudinal stripes on a yellow or blue background. Signals are provided by certain postures which display the stripe and colouration, as shown in figure 154. Tactile signals may be given by the 'cleaner' wriggling over the flank of the 'host'. These signals serve to quieten active hosts which normally swim vigorously and, of course, help to identify the cleaners and deflect the feeding behaviour of the hosts.

Alarm signals such as the chattering of blackbirds and the stamping of rabbits may also be recognised as such by other species in the area.

5.6.1 Communication in honey-bees

Reference has already been made (in section 5.3.2) to the work of von Frisch on communication between honey-bees.

SAQ 150 From the passage quoted, say what information was given by the scout bees to the rest of the hive.

SAQ 151 What method of communication was used?

Normally, the returning bees dance on vertical combs within a dark hive. Figure 155(*a*) shows the round dance of the bee used when the food source is close to the hive. Other bees track the dancer's side with their antennae. This dance contains no directional information. New foragers seek in all directions around the hive.

155   Dances of the honey-bee:
(*a*) round dance;

(*b*) waggle dance

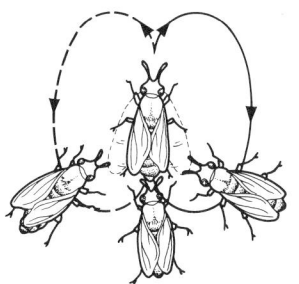

When the food source is more than 70 m from the hive, the bee waggle-dances and her abdomen traces a figure 8, as shown in figure 155(*b*). The bees closer to the dancer perceive the odour of the blossoms visited by her and learn the distance and direction they have to search. The speed of this dance is related to the distance of the food source from the hive, or rather the energy that the bees will expend in reaching their goal (thus compensating for head-winds, and so on). While performing the straight component of the dance, the bee produces a rasping sound with her wings and sounds are part of the communication. Figure 156 demonstrates how the bee dances on the comb to indicate a given location of feeding place.

156   Direction indication according to the position of the Sun in the honey-bee waggle dance

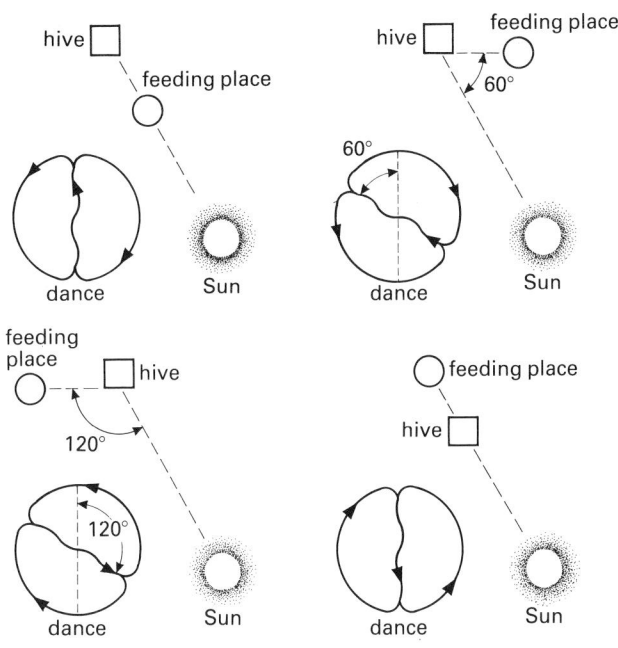

SAQ 152 Study figure 156 and explain briefly how the orientation of the dance conveys directional information to the other bees.

5.6.2 Pheromones

Chemical systems provide the dominant means of communication in many animal species. Even animals with well-developed vision and hearing may use chemical identification for food and mates. A pheromone is defined as being a chemical substance which influences the behaviour or development of members of the same species. It may be considered to have a 'regulating' effect on an animal's external environment, as hormones do for the internal environment.

Pheromones have particular importance among social insects, such as ants, which tend to live in the dark. The pheromone vocabulary of ants includes short-lived chemical trails that guide workers to a food source, alarm substances (again, short-lived) that can stimulate a colony into instant action, and secretions that encourage grooming, food exchange, care of the queen and the young, and so on.

The ability of a minute amount of pheromone produced by a female moth to attract males is almost unbelievable. It has been calculated that 0.01 µg of gyplure, the attractant of the gypsy moth, is theoretically capable of attracting more than a billion male moths! The pheromone enables the female to advertise her presence over a large area with a minimum expenditure of energy. Male moths detect the pheromone with their broad and finely divided antennae. They seem to fly upwind and thus towards the female; as they approach the female there is a slight increase in chemical strength and this guides them over the remaining distance.

Chemical signals are also important in mammals and are involved in territorial marking, sexual behaviour and social identification. It has been discovered that the odour of a male mouse can initiate and synchronise the oestrous cycles of female mice. The new pregnancy of a female mouse can be stopped by presenting the odour of a strange male. Possibly, the strange male odour suppresses secretion of prolactin.

SAQ 153 What is the normal effect of prolactin?

5.6.3 Song

Both bird and locust males use courtship songs in order to attract partners over great distances. There is evidence that the song of some males is more attractive to females than that of other males of the same species. The 'successful' bird has the more complex song.

Bird song plays a part in species recognition and also in territorial defence. Male robins use prominent perches at the boundaries of their territory for a song. Birds also produce alarm and warning calls.

Behaviourists have generally directed their attention to the nature of the development of species-specific song and less is known about the precise information communicated by song. In general, song conveys information about species, sex, location and motivation. The song of male insects, such as

crickets and grasshoppers, is produced by interaction between wings or legs. Information is coded as intensity and frequency. If a sexually receptive female hears a stationary male, she moves rapidly towards him. After several precise recognition signals, the male lowers his wings and produces another type of call. This stimulates the female to mount the male.

5.6.4 Courtship

Courtship may be considered as a form of communication, a signal system for the transfer of important pieces of information between potential mating partners. Usually, the success of this information involves a variety of signal types presented in a definite order and combination. Colour, form and movement are involved in the displays of many male birds. Females of many bird species signal their readiness to mate by a low crouching posture and visual signals are used also by fish, mammals and arthropods. Auditory signals were mentioned in the last section; they are also important to frogs and toads. Tactile signals are used by many species, particularly reptiles, newts and salamanders and have great importance in human courtship. The role of chemical signals in courtship has also been referred to.

Figure 157 shows a diagrammatic and a schematic representation of courtship behaviour in the three-spined stickleback, as known from the work of Tinbergen. You can read for yourself the account of the investigation of this behaviour in the *Scientific American* offprint, *The curious behaviour of the stickleback*. The pattern of courtship revealed here illustrates the combination of visual, tactile and possibly chemical stimuli and also the reciprocal nature of the information exchange, each act presenting a sign stimulus to the other sex. This sequence might be considered as a **chain reaction;** each small piece of stereotyped behaviour providing the stimulus for the next one.

Stickleback behaviour illustrates well the functions of courtship: it serves to bring the sexes together. The normally aggressive behaviour of the territorial male is temporarily suppressed and the reproductive behaviour of the pair is synchronised so that egg deposition and subsequent fertilisation by the sperm occur in close time. The highly specific signal exchange also ensures **reproductive isolation**, that is breeding will only take place between members of the same species.

SAQ 154 What other advantage does the courtship pattern of the three-spined stickleback confer?

157 The courtship of the three-spined stickleback

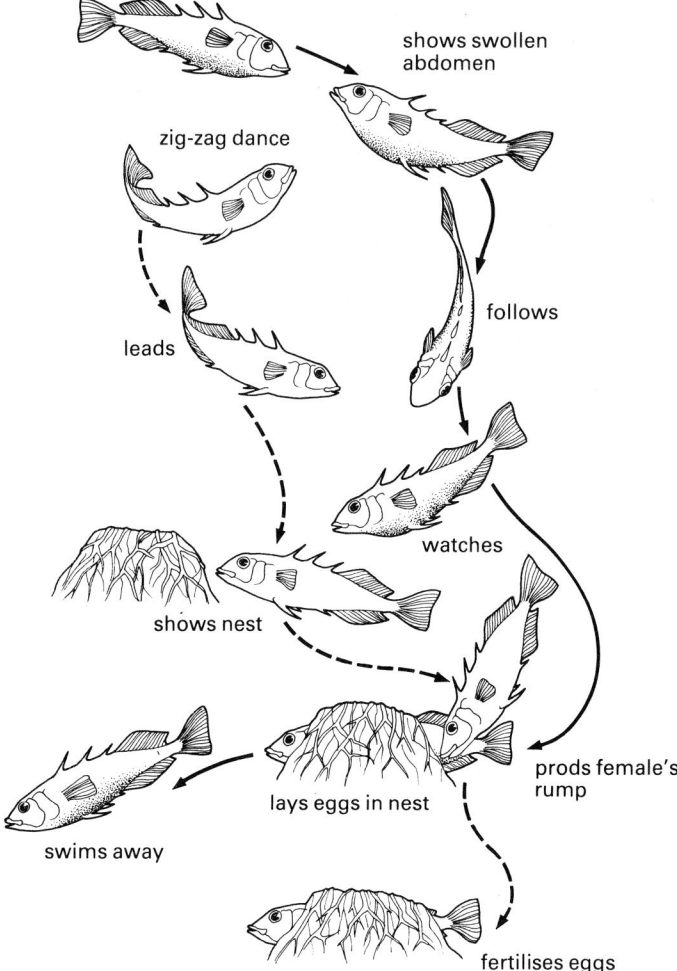

zig-zag dance

shows swollen abdomen

follows

leads

watches

shows nest

prods female's rump

swims away

lays eggs in nest

fertilises eggs

5.6.5 Extension: The curious behaviour of the stickleback re-examined

For almost 30 years, the account of Tinbergen of the reproduction and courtship in the three-spined stickleback has been cited as an example of deductive science at its best. Read for yourself *Scientific American* offprint 419 (Dec. 1952) *The curious behaviour of the stickleback* by N. Tinbergen. Recently, however, a number of papers have been published which seriously question some aspects of Tinbergen's findings.

Consider this statement of Tinbergen: "We found that the red models were always more provoking than the others, though even the silvery or green intruders caused some hostility." Rowland (1982) carefully repeated Tinbergen's experiment but found that adding red colour to a model made it less likely to be attacked by a territorial male. Grey models received a statistically significant greater number of attacks than red models.

Other workers have discovered that females too may be very aggressive when in reproductive condition, a fact not mentioned by Tinbergen. They have also discovered that the presence of eggs in a male's nest increases the chance that a female will spawn. This throws light on what Tinbergen describes as a 'displacement activity' fanning an empty nest.

SAQ 155 Suggest a function for this fanning in the light of the fact that the presence of eggs in a nest increases the chance that a female will spawn.

Your tutor may be able to supply you with further details of these recent studies. Even this brief outline should draw your attention to three points.

1 Even a recognised and much-acclaimed piece of research may not necessarily be correct (or repeatable) in every detail.

2 More detailed observations may still be made and new insights gained.

3 The way scientists view the world is influenced by the way earlier scientists have seen it. For example, other people had tried to repeat the investigation with red dummies and failed to repeat Tinbergen's results. They tended to assume that in some way they were in error! (This same point was made with reference to the classic work on tropisms in the unit *Response to the environment*.)

5.7 Social behaviour

Social behaviour involves two or more animals, more usually two members of the same species. You have already met numerous examples of social behaviour in this section, all involving communication in some form. Courtship is an example of a sustained period of social behaviour for some species. Some animals are largely solitary, except for the necessary contact for reproduction and rearing of young. Many animal species live in some kind of grouping:

– a pride of lions,   – a troop of baboons,

– a herd of deer,   – a nest of ants or termites,

and so on.

Some animal groups are described as **societies**. They do more than merely aggregate, they influence one another; there is a degree of cooperation and communication between members. Their activity may be synchronised and there is a means of recognising the members of that particular group (such as visual, chemical).

Social life confers certain advantages on its members:

– it may offer protection from predators,

– it may increase feeding efficiency and reproductive efficiency,

– division of labour may contribute to all of these.

However, social life may also increase *competition* for food, mates and so on, and therefore there may be a 'trade-off' between advantages and disadvantages, particularly for less-successful or low-ranking individuals.

Limitations of space in this unit and, possibly, time in your course, mean that this topic will not be further developed here. However, references to this topic can be found at the end of this section.

5.8 The evolution of behaviour

In the introduction to this section, four kinds of question were asked about behaviour. Three of these questions have received attention but very little has been said in answer to the question 'How did this behaviour evolve?'

In order to answer this question, a comparative approach must be adopted. An example of a very broad comparative sweep across the animal phyla gave the evolutionary trend shown in figure 139 which is linked to the trend for complexity in the evolution of nervous systems. Other comparisons may be made on a somewhat narrower basis. Behaviour can evolve only if it *varies* and if such variations are transmitted from one generation to the next. Such variations are usually produced by *genetic* changes involving new *mutations* or *recombinations* of genes already present. Most human behaviour is *culturally transmitted* from generation to generation and examples of this evolutionary phenomenon are fairly common in primates but occur also in other mammals and birds. However, they are considered by some to play only a very limited role in the evolution of behaviour, though this is currently an area of debate.

5.8.1 Cross-species studies

A comparative study of vertebrates with limbs reveals similarities in the way a dog or cat scratches its face, a bird preens its head feathers, and a lizard scratches its skull. All scratch with the hindlimb and form a sort of tripod for support, as can be seen in figure 158. In the case of birds it seems clumsy and unnecessary for the wings to be lowered. One explanation is that the scratch reflex is inborn and has not changed as rapidly as the wings have evolved.

Several studies have been made on closely related groups of birds, such as gulls, geese and ducks. There are nine species of lovebird in the genus *Agapornis*. Their closest living relatives are the hanging parakeets of Asia. On grounds of pair-colouring and social organisation, the Madagascar lovebird, the Abyssinian lovebird and the red-faced lovebird are close to the ancestral form, while the remaining species are more divergent. W.C. Dilger has made a detailed laboratory-based study of these birds at Cornell University. The table

158   Mammal and bird scratching

159   Behaviour patterns of lovebirds (*Agapornis*) (data from Dilger)

	Courtship feeding	Accompanying head bobs	Nest materials	Method of carrying nest material	Form of nest	Cavity defense display
Madagascar lovebird (*A. cana*)	males and females feed each other	rapid, numerous, trace a small arc followed by prolonged bill contacts while food transferred	very small pieces of bark and leaves deposited on floor of a cavity	tuck materials into body feathers (all over)	unshaped deposits of material	feather ruffling by female, wings and tail spread, harsh sound, will lunge at intruder
Peach-faced lovebird (*A. roseicollis*)	males feed females only, females fluff feathers	head bobbings slower, fewer, trace wider arc, bill contacts for short time only	longer strips of bark and leaves (within cavity)	tuck into feathers of lower back only, very rarely carry in bill	cup-shaped nest	'mobbing' by other birds of colony before cavity is reached
Fischer's lovebird (*A.personata-Fischeri*)	males feed females who adopt a special position	much as for peach faced but slower and wider	twigs as well as strips of bark and leaves (within cavity)	carry nest materials in bill, one piece at a time	elaborate covered nest entered through tunnel	as above

in figure 159 compares six behaviour patterns across three species based on Dilger's findings.

Hybrid offspring of peach-faced and Fischer's lovebird display a conflict in behaviour between the tendency to carry nesting material in bill or feathers.

Tucking behaviour persisted even when they finally carried material in their bills. After three years of experience they carried in their bills and rarely showed tucking behaviour (though this was more efficiently performed when it occurred – see figure 160).

160 Hybrid lovebird carrying nesting material

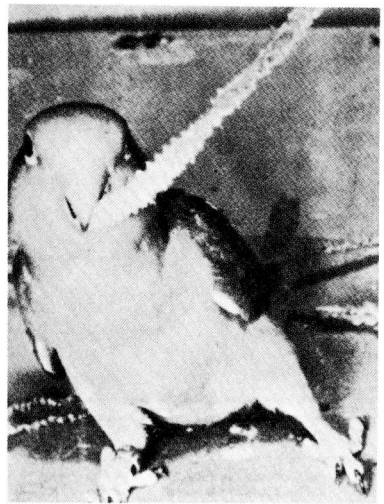

SAQ 156 (*a*) Describe the evolutionary trend in the carrying of nest materials as revealed by figure 159. (*b*) How does the behaviour of the hybrid support the evidence from the comparative study?

SAQ 157 What light does the answer to SAQ 156 throw on the issue of instinctive and learned behaviour?

SAQ 158 Why do you think the Madagascar lovebird is more aggressive than the peach-faced and Fischer's lovebirds?

5.8.2 Genes and behaviour

The fruit fly *Drosophila* is better known genetically than any other animal. Hundreds of mutant genes are known and their position on the chromosomes has been mapped (see unit *Genetics*). The genes *vestigial* and *dumpy* alter the form of the wings and another mutation, *yellow*, produces flies with yellow bodies. However, these genes have effects other than the one for which they are named, in particular the males possessing them are less successful than normal males in stimulating females to mate with them.

Drosophila courtship involves three phases described as orientation, vibration and licking. Orientation involves the approach and following of the female by the male. Following this, vibration is the wing display of the male which stimulates sense organs at the base of the female's antennae. Licking is the final stage of the courtship when the male goes behind the female, licks the genital area with his proboscis and attempts to copulate.

Figure 161 is a table analysing courtship of wild-type and yellow males with wild-type females in *D. melanogaster*.

SAQ 159 Study figure 161 and list three ways in which the courtship behaviour of yellow males differs from that of wild-type males.

SAQ 160 Give a possible reason for the lower success rate in courtship for vestigial and dumpy males.

The examples of lovebirds and fruit flies indicate some of the ways in which research into behaviour is trying to answer the questions relating to the

161 Analysis of courtship of wild and yellow males with wild females in *Drosophila melanogaster* (data from Bastock, 1956)

	No. of records out of 100 in which courtship activities were shown	Courtship records			Bout length (in units of 1.5 s)	
		Orientation (O)	Vibration (V)	Licking (L)	Average (V + L)	(O)
wild-type ♀ × wild-type ♂	92	72	22	6	3.9	5.5
wild-type ♀ × yellow ♂	83	77	18	6	2.9	6.9

evolution of behaviour and the genetic control of behaviour.

5.9 A case-study in behaviour: the male Siamese fighting fish

The Siamese fighting fish, *Betta splendens*, shown in figure 162, is sometimes kept in aquaria in Far Eastern countries so that bets can be laid on the outcome of fights between rival males. In the breeding season, the males become very aggressive and advertise their right to a particular territory by a brilliant metallic red and blue colouration. When a female approaches, this colourful display serves to bring her to a state of readiness for egg-laying. The male prepares a nest formed from bubbles of air blown from sticky mucus.

162   Siamese fighting fish

AV 3 shows sequences of the behaviour of Siamese fighting fish for you to observe and offers you the opportunity to construct a partial ethogram of the fish's behaviour. You can watch investigations into the nature of the visual stimuli which elicit the aggressive display of the males and see the results of attempts to measure and quantify the display.

AV 3: The display of the Siamese fighting fish

Materials

VCR and monitor
ABAL video sequence: *The display of the Siamese fighting fish*
Worksheets

Procedure

(*a*) Check that you have all the relevant materials for this activity.

(*b*) Check that the video cassette is set up ready to show the appropriate sequence – *The display of the Siamese fighting fish*.

(*c*) Start the VCR and stop it to complete the worksheets as indicated in the film.

(*d*) If you do not understand anything, stop the video, rewind, and study the relevant material again before consulting your tutor.

(*e*) If possible, work through the video and worksheets with a small group and discuss the material with your fellow students.

5.10 Summary assignment 9

1 Make patterned notes to explain the following terms in relation to behaviour:
   maturation, motivation, hormones, pheromones and song.
Provide an example for each term either in your notes or separately.

2 Draw up a table and for each of the terms below, (*a*) write a definition, (*b*) give an example, and (*c*) explain its biological significance.
   endogenous rhythms, photoperiodism in animals, the dance of the honey-bees, courtship, social behaviour

3(*a*) Write a concise paragraph to describe the possible evolution of nest-building behaviour in lovebirds.

(*b*) Explain briefly, with examples, how the change of a single gene may affect mating behaviour in *Drosophila*.

Self test 7 on page 115 covers sections 5.5–5.8 of this unit.

5.11 Past examination questions

1 In a laboratory experiment a number of woodlice (*Porcellio scaber*) were kept under identical conditions. Eight choice chambers were set up, each containing a different and known percentage relative humidity. A woodlouse was placed in each chamber and the distance moved, the number of turns made and the time spent at rest were measured in each case. This was repeated and the averages of the results for movement and turning are given in the table shown in figure 163.

163   Behaviour of woodlice in a choice chamber

Chamber	1	2	3	4	5	6	7	8
% RH	10	20	50	60	70	80	90	100
Centimetres moved per minute	22.5	22.5	21.5	21.5	20.0	16.5	11.5	9
Number of turns per hour	62	54	50	45	40	32	25	20

(*a*) (i) Draw a graph to display the data shown in figure 163.
(ii) On the graph draw the line you would expect for the time spent at rest in each percentage relative humidity.
(iii) Name the behavioural terms used for the two orientation behaviours involved.
(iv) Under what conditions would the results suggest the woodlice were kept before the experiment? *Briefly* explain your answer.

(*b*) (i) Name a substance or substances which could be used to produce the known percentage relative humidities.
(ii) Name *one* variable, other than temperature, which could interfere with the woodlouse response to humidity and explain *briefly* the effect it could have.
(iii) What precaution(s) would you propose to remove any effect of this variable?
(iv) Why should only one woodlouse at a time be used in a choice chamber?

(*c*) Describe how the responses displayed have survival value for woodlice in natural situations.
(JMB, 1982)

2 Write an essay on the influence of hormones on animal behaviour.
(JMB, 1981)

3(*a*) What is meant by reinforcement in animal learning?
(*b*) Describe *two* examples to show how reinforcement may be used in the training of animals.
(*c*) Give, with brief explanations, examples of learning which do not require reinforcement.
(*d*) Explain what is meant by 'insight learning'.
(JMB, 1981)

4 How do animals other than man communicate with each other and what do they communicate?
(JMB, 1983)

5(*a*) (i) Briefly describe *six* functions of courtship.
(ii) Describe the sequence of stimuli and responses shown in courtship in *one named* example of an invertebrate or fish.
(*b*) Describe an innate sequence of stimuli and responses which occur in a situation other than courtship for any *one named* animal.
(JMB, 1983)

6 Male zebra finches show courtship behaviour towards female zebra finches. Experiments were performed in each of which a male zebra finch was caged with (i) a female zebra finch with a red beak, (ii) a female zebra finch with a black beak, (iii) a simple model of a female zebra finch with a grey beak.

The results obtained are expressed in the histograms shown in figure 164.

164   Courtship behaviour in zebra finches

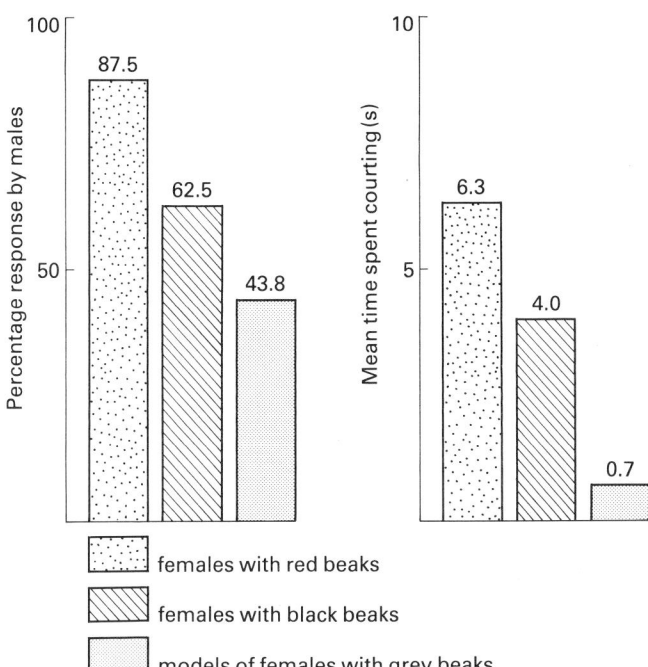

females with red beaks

females with black beaks

models of females with grey beaks

(*a*) Comment on and explain these results as fully as you can.

(*b*) Explain the significance of the performance by the males of apparently meaningless acts, such as preening, when caged with models of females. Give *one* other example of this form of behaviour.
(London, 1978)

5.12 Recommended reading

Psychobiology: the biological basis of behaviour by J.L. McGaugh, N.M. Weinberger and R.E. Whalen – this book of readings from *Scientific American* contains a number of accounts which have been referred to in section 5:
The evolution of intelligence by M.E. Bitterman.
The behaviour of lovebirds by W.C. Dilger.
Imprinting in animals by E.H. Hess.
Love in infant monkeys by H.F. Harlow.
Learning to think by H.F. Harlow & M.R. Harlow.
The curious behaviour of the stickleback by N. Tinbergen.

Other articles in this book are of interest and will develop your knowledge and understanding of behaviour.

The study of behaviour by J.D. Carthy, Studies in Biology No.3.

Chemical communication by J. Ebling & K.C. Highnam, Studies in Biology No.19.

Section 6 Self tests

Self test 1

1 What is the difference between movement and locomotion?

2 State four reasons why movement is important to animals.

3 List three different types of plant movement.

4 Give three reasons why organisms must be supported.

5 Name three different types of skeleton and give an example of each.

6 List four functions of a skeleton.

7 Name four principles or factors involved in the movement of animals.

8 Why do functional muscles usually occur in pairs?

9 What name is given to such pairs of muscles?

10 What are the three main materials involved in plant support?

Self test 2

1 Give the chemical nature of the following substances: (*a*) collagen, (*b*) sclerotin, (*c*) chitin, (*d*) cellulose, (*e*) lignin.
For each of these substances say where it can be found acting as a support material.

2 State three common features of all connective tissues.

3 What type of connective tissue is found in (*a*) tendons, (*b*) mesenteries of the abdomen, (*c*) fat around kidneys.

4 Distinguish between hyaline cartilage, fibrocartilage and elastic cartilage. Indicate a place where each may be found in a mammal.

5 Draw a diagram of a long bone. Indicate the following structures on your diagram:
   compact and spongy bone, epiphyseal cartilage, regions of bone growth, marrow cavity, periosteum.

6(*a*) What are the names given to the cell membrane and the cytoplasm in the muscle fibre?
(*b*) Name four structures found in the cytoplasm.

7 During muscle contraction, what happens to (*a*) sarcomere length, (*b*) length of actin and myosin filaments, (*c*) A band, (*d*) I band.

8 What is the name given to the region of association between a nerve ending and a muscle fibre?

9 Briefly say what happens when an action potential arrives at the end of a nerve fibre.

10(*a*) Name the cells shown in figure 165.

165   Plant cells

(a)
(b)

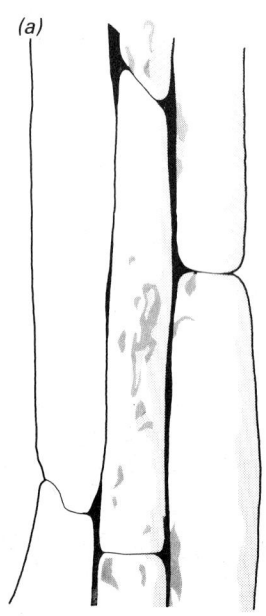
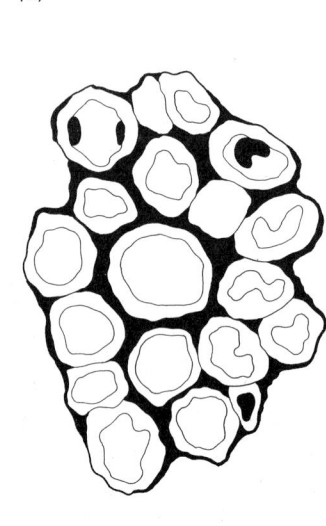

(*b*) State the major component of the wall of each cell.

(*c*) State two ways in which you could identify these cells in sections under the microscope.

(*d*) In what parts of the stem are these two types of cell most frequently found?

11(*a*) State the two major components of the procuticle of an insect.

(*b*) What additional substance is found in the epicuticle?

(*c*) How does the cuticle of an insect differ at the joints from other parts of the cuticle?

12 Explain the two major disadvantages of an exoskeleton.

Self test 3

1(*a*) State the four types of force which may affect organisms.

(*b*) List six factors associated with the organism and its environment which may cause these forces.

2(*a*) What is meant by the centre of gravity?

(*b*) How will this affect stability in animals **A** and **B** in figure 166?

3 In what way is the positioning of limbs in mammals more efficient than that in reptiles?

4 How does the distribution of compact bone and the arrangement of trabeculae aid support?

5 Explain the role played in support by muscles **A** and **B** in figure 167.

167 Muscles of the hip–thigh region

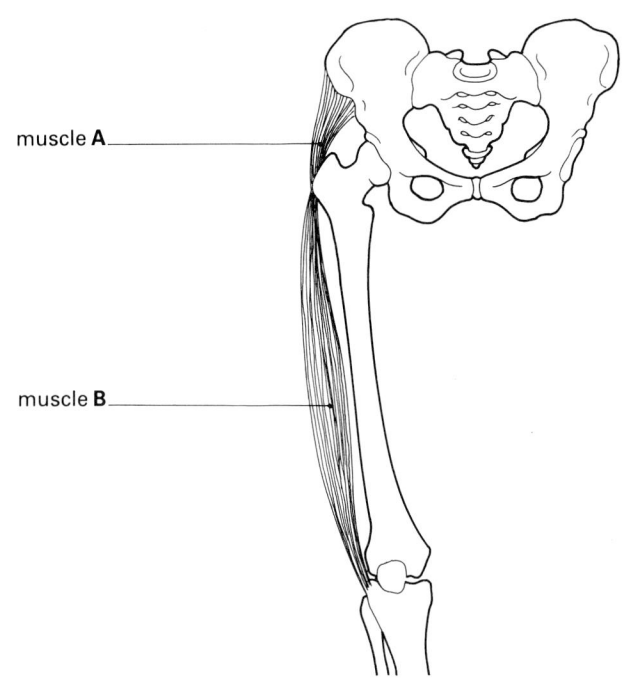

muscle **A**

muscle **B**

166 Two animals

6 In the wallaby there is a single cantilever girder associated with the hindlimbs (see figure 168).
(*a*) What does this mean?
(*b*) What is the significance of this for the animal?

168   Skeleton of a wallaby

7 What role do the head and neck of a giraffe play in maintaining support?

8 Examine figure 169. Explain one way in which the design of the foot is adapted to support the mass of the body.

169   Structure of the foot

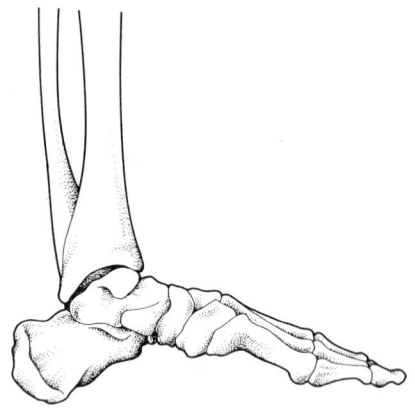

9 Write down the names of labels (*a*)–(*j*) on the diagram of a mammalian skeleton shown in figure 170.

170   A mammalian skeleton

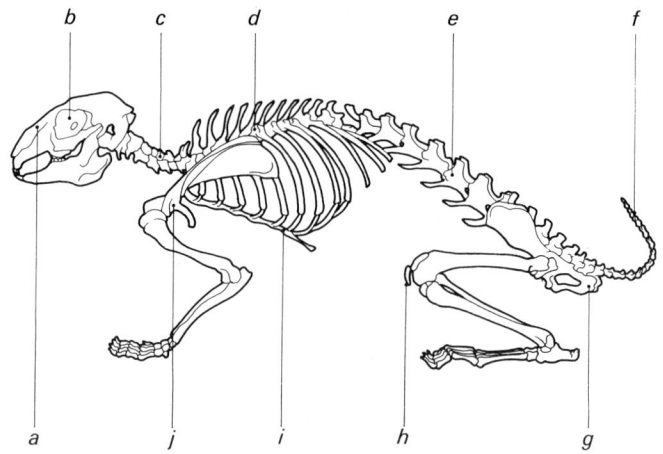

10 Identify and write down the names of labels (*a*)–(*d*) of the bone shown in figure 171.

171   Mammalian bone

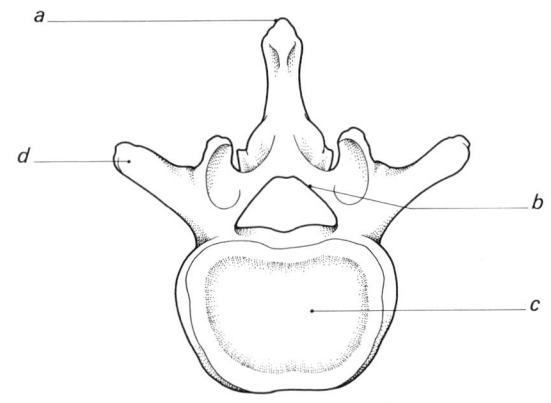

Self test 4

1 Examine figure 172.
(*a*) Identify limbs **A** and **B**.
(*b*) How does each of these limbs differ from the basic pentadactyl plan?

172 Mammalian forelimbs

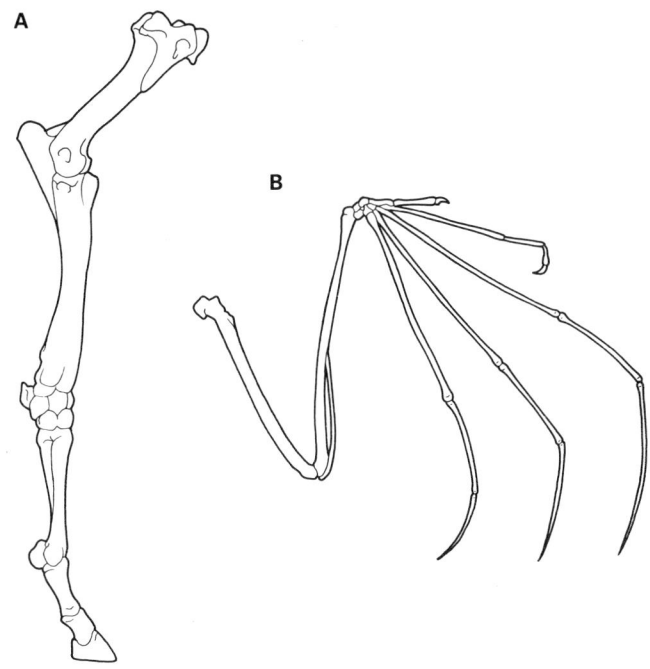

2 For the three main types of joint
(*a*) state what structures are involved;
(*b*) what degree of movement is possible?

3 State as fully as possible what type of joint is found in each of the following.
(*a*) Joint between shoulder blade and collar bone
(*b*) Joint between humerus and scapula
(*c*) Joint at knee
(*d*) Joint between two thoracic vertebrae
(*e*) Joint between a carpal and a metacarpal

4 What is the function of the following in movement of a mammalian limb?
(*a*) Long bones   (*b*) Muscles   (*c*) Joints (*d*) Nerves

5 How are proprioceptors important in support?

6 Explain how antagonistic muscle pairs bring about bending and straightening of the limb with reference to the human leg.

7 If a frog's gastrocnemius muscle is dissected out, it can be stimulated electrically and the effects recorded by means of a kymograph. Make sketch diagrams to show the kymograph traces you would expect in the following situations:
(*a*) a single stimulus,
(*b*) a series of stimuli with a frequency of 8 per second,
(*c*) a series of stimuli with a frequency of 25 per second.

8 Explain the following terms in relation to muscle action:
    twitch, summation, tetanus, fatigue, all-or-nothing response, refractory period.

9 Figure 173 shows the deltoid muscle which is involved in raising the arm.
(*a*) Sketch the diagram and indicate on your diagram the position of the fulcrum, effort and load.
(*b*) What class of lever does this represent?
(*c*) Does it have a high or low mechanical advantage?
(*d*) What is the significance of your answer?

173   Muscle involved in raising the arm

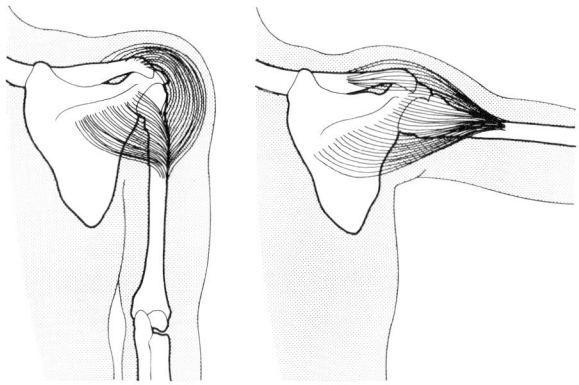

10 What is the function of a tendon?

11 Outline the role of (*a*) the cerebellum and (*b*) the cerebral hemispheres in the control of movement.

12 Draw an outline sketch of a tree and on it indicate clearly the forces acting on such a tree and their source.

Self test 5

1(*a*) Explain the term amoeboid movement.
(*b*) What changes occur in the cytoplasm during this movement?

2 Compare locomotion in *Paramecium* and *Euglena*.

3 What are the roles of the following in an earthworm: body wall, coelom, chaetae?

4 Explain the function of the muscles labelled **A–D** in figure 174 during walking in human beings.

174   Muscles involved in walking in human beings

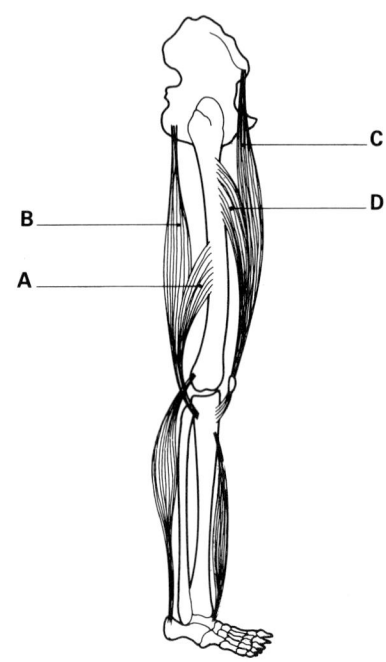

5 Explain the role of the following in locomotion in a fish:

myotomes, tail, vertical fins, horizontal fins, swim bladder of teleosts.

6 Draw a diagram to explain the main forces acting on a bird's wing during flight.

7 During flight in birds
(*a*) which factors affect lift?
(*b*) which factors affect speed of gliding?

8 During slow flapping flight in birds
(*a*) which stroke is mainly associated with lift?
(*b*) which stroke is mainly associated with forward propulsion?

9 State six features of a bird's body which adapt it for flight.

10 Briefly explain how flight is achieved in an insect. Use diagrams to illustrate your answer.

Self test 6

1(*a*) Say briefly why Konrad Lorenz is regarded as a key figure in the development of ethology.
(*b*) What is most characteristic of his method of studying animals?
(*c*) Name one area of behaviour to which he made a major contribution.

2 List five techniques used to supplement unaided observation in the study of behaviour and indicate the type of information revealed by each of them.

3 Briefly outline the value of models in behavioural investigation.

4 Explain the following terms and give an example to illustrate each of them:

fixed action pattern, sign stimulus/releaser, supernormal stimulus, reflexes.

5(*a*) State four differences between innate and learned behaviour.
(*b*) What is the danger in using such terms?

6 What kinds of behaviour are illustrated by the following examples?
(*a*) A student trying to solve a maths problem suddenly 'sees' the solution.
(*b*) Chemicals diffusing from certain organisms cause an *Amoeba* to move towards them.
(*c*) If a hungry rat is placed in a maze and rewarded with food at the end, it makes many mistakes by going down blind alleys before reaching the goal, that is the food. If a rat which is not hungry/thirsty and which had previously been allowed to explore the maze is placed in the maze, it makes far fewer mistakes before reaching the goal.

(*d*) If the anterior end of an earthworm is touched, it immediately withdraws. If the earthworm is repeatedly touched, it will soon cease to withdraw.

(*e*) A cat is placed in a cage which can be opened from the inside by depressing a lever. The cat moves around the cage and eventually steps on the lever by chance and releases itself. This experiment is repeated several times. Eventually, the cat moves to the lever and depresses it as soon as it is confined.

(*f*) Woodlice move faster where the humidity is low and slow down as humidity increases.

7 Distinguish between a kinesis and a taxis.

8 What is meant by imprinting? How does this differ from other forms of learning?

9 Explain the difference between classical and operant conditioning.

10 How may a 'learning set' be formed? Why is it useful?

Self test 7

1 What role does the hypothalamus play in motivation?

2 Explain the connection between the photoperiod, hormones and reproductive behaviour in some birds and mammals.

3 Name three types of endogenous rhythm, and give an example of each.

4 Study figure 175 and state the purpose of these movements of a worker bee.

175 Movements of a worker bee in a hive

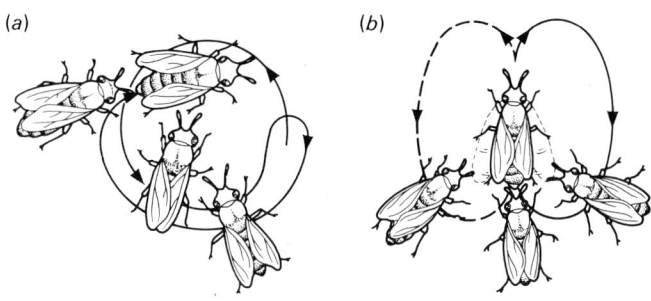

(*a*)　　　　　　　　(*b*)

5 What are pheromones? Give two examples of behaviour connected with pheromones.

6 Give three functions of bird song.

7 What is the biological purpose of courtship?

8 Give three advantages of social life for animals.

9 The nest-building behaviour of lovebirds of the genus *Agapornis* reveals an evolutionary trend. What is this trend? Why were the hybrid offspring of two of the species particularly interesting in their behaviour?

10 Explain how the genes for (*a*) bar eye and (*b*) dumpy wings may also affect the reproductive success of *Drosophila*. How does this knowledge relate to understanding of the evolution of behaviour?

Section 7 Answers to self tests

Self test 1

1 Movement covers the change in position of an organism or a part of an organism. Locomotion is a specific type of movement and refers to the movement of the whole body from one place to another.

2 Any four of the following.
Jaw movements to break up food,
Avoidance of unfavourable conditions, such as lack of food or extreme temperature
Avoidance of predators
Gathering food
Facial movements convey emotions/information
Finding a mate

You may have thought of other reasons also.

3 Tropic movements, nastic movements, sleep movements, trapping movements of insectivorous plants

4 Provides definite shape.
Enables organism to exert force against medium.
Enables organism to react against other forces.

5 Exoskeleton – the 'shell' of a crab or other crustaceans, or the cuticle of insects or the test of a sea-urchin
Endoskeleton – the 'bones' of vertebrates (or cartilage)
Hydrostatic skeleton – coelomic fluid of earthworm, turgor pressure of living plant cells, fluid in enteron of *Hydra*, etc.

6 Support for 'body', protection of delicate organs, firm anchorage for muscles, levers for movement

7 Need to overcome friction and resistance of the medium.
Energy needed: potential energy → kinetic or mechanical energy
Movement requires exertion of force against the medium.
Moving parts associated with contractile tissue.
Control mechanism for coordination of activities.

8 Because muscles can actively contract but an external force must be used to restore the original length quickly. This force is supplied by the contraction of the second muscle of the pair (and vice versa).

9 Antagonistic pairs, circular and longitudinal muscles

10 Water (turgor pressure – 'hydrostatic skeletons')
Cellulose in secondary cell walls
Lignin or wood in xylem tissue

Self test 2

1(*a*) Protein: present in bone and cartilage
(*b*) Tanned protein: cuticle of insects and arachnids.
(*c*) Nitrogen-containing polysaccharide: cuticle of insects
(*d*) Polysaccharide: secondary wall of plants cells, such as fibres
(*e*) Complex aromatic compound: thickened walls of xylem tissue

2 Matrix of non-living material
Cells which secrete the matrix
Fibres running in the matrix

3(*a*) Dense fibrous connective tissue
(*b*) Areolar connective tissue
(*c*) Adipose tissue

4 Hyaline cartilage does not contain yellow elastic fibres; found in bones of embryo and young mammals.
Fibrocartilage contains a very large number of collagen fibres; found in invertebral discs.
In yellow elastic cartilage elastic fibres predominate; found in the pinna of the ear.

5 See figure 176.

176 Diagram of a long bone

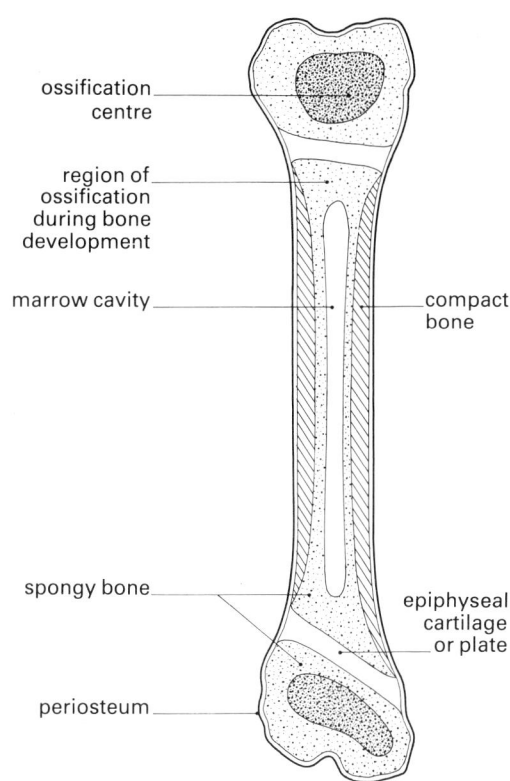

ossification centre

region of ossification during bone development

marrow cavity

compact bone

spongy bone

epiphyseal cartilage or plate

periosteum

6(*a*) Membrane – sarcolemma; cytoplasm – sarcoplasm
(*b*) Mitochondria, sarcoplasmic reticulum, T-system, myofibrils

7(*a*) Decreases
(*b*) Remains the same
(*c*) Remains the same
(*d*) Gets shorter

8 Motor end-plate

9 Acetylcholine is released into the space between the nerve and muscle membrane. Acetylcholine reacts with receptor molecules on the sarcolemma of the muscle fibre. This causes the latter to become highly permeable to small cations. Sodium ions flow into the muscle fibre and the membrane is depolarised. Thus, an action potential is set up. This spreads over the surface of the muscle fibre through the T-system and through the sarcoplasmic reticulum. This stimulates the release of calcium ions into the sarcoplasm which, in turn, removes the inhibitory effect of troponin and tropomyosin. Myosin is thus available to catalyse the breakdown of ATP and the release of energy.

10(*a*) Cell **A** – collenchyma; cell **B** – sclerenchyma
(*b*) Collenchyma – cellulose; sclerenchyma – lignin
(*c*) (i) Degree and regularity of thickness of cell walls
(ii) Reactions to specific microscopical stains.
(*d*) Collenchyma – often found as a cylinder just within the epidermis.
Sclerenchyma – found as cylinders within the cortex of roots or stems and associated with vascular bundles.

11(*a*) Chitin fibres and a protein matrix
(*b*) Wax
(*c*) The cuticle is much thinner at the joints. It comprises endocuticle with no exocuticle or epicuticle.

12(i) The exoskeleton is unable to grow. This means moulting is necessary during growth of the insect. Between moults, the insect is very vulnerable to attack and is without an adequate support system.
(ii) For animals with an exoskeleton, as overall size increases, the surface area and thickness of the exoskeleton would have to increase greatly to provide the necessary support and protection for a larger animal. This would result in increased mass and reduced mobility and lack of flexibility. This is one reason why insects have never reached a very great size.

Self test 3

1(*a*) Push/compression forces, pulling/tension forces, sliding/shearing forces, twisting/torsion forces
(*b*) Six from the following.
Mass of animal, force of gravity, movement of the animal itself, wind currents, water currents, carrying young, fighting

2(*a*) The centre of gravity is the point in the body at which the weight of the body can be considered to be concentrated. If the body is tilted so the centre of gravity comes to lie outside the base, the body becomes unstable and will be liable to fall.
(*b*) Animal **A** has a higher centre of gravity than animal **B**. Animal **A** will become unstable more quickly than animal **B** when both are tilted.

3 In mammals, the limbs are positioned under the body. In this way they are able to support the body off the ground with a minimum muscular effort, reduced need for a large pelvic girdle and minimum expenditure of energy.

4 Bones are subject to maximum stress at the periphery of the shaft and at the ends. The periphery of the shaft is made of compact bone which is stronger than spongy bone.

Maximum resistance to stress is obtained if the trabeculae are arranged corresponding to the lines of maximum strain (trajectories). This occurs in the ends of long bones. Stress here comes from several directions. The trabeculae are aligned to correspond with these directions. The trabeculae can change if the forces acting on the bone change.

5 Muscles **A** and **B** act as braces to reduce the stress on the femur caused by the mass of the body acting asymmetrically down on it.

6(a) A cantilever girder is supported by an upright strut at one end only. The girder is composed of two main members which oppose the two main forces acting on it. These are the tension member and the compression member.
In the wallaby, the hindlimb acts as a strut. The vertebral column forms a bracket or girder attached to the strut and supports the body (see figure 177). The centra of the vertebrae act as the main compression member. The neural spines act as secondary compression members. Muscles and ligaments act as tension members.
(b) The mass of the animal is centred over the hindlimbs which is where its centre of gravity lies.

177   The wallaby skeleton as a cantilever system

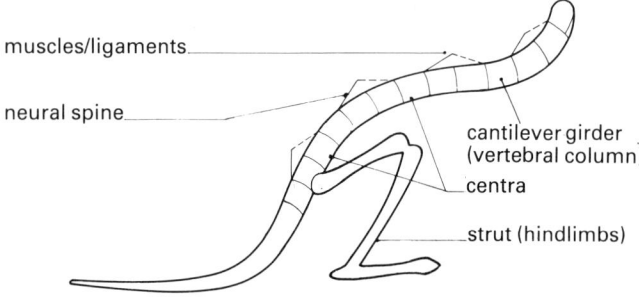

muscles/ligaments
neural spine
cantilever girder (vertebral column)
centra
strut (hindlimbs)

7 The main mass of the giraffe's body is centred on the forelimbs. The head acts as a counterbalance against the mass of the rest of the body.

8 Arches provide much more support than flat surfaces. The human foot is designed so that it is composed of two arches. These are stronger and provide better support and stability for the mass of the body than flattened structures would do.

9(a) Skull   (b) Orbit   (c) Cervical vertebra   (d) Thoracic vertebra   (e) Lumbar vertebra   (f) Caudal vertebra   (g) Pelvic girdle   (h) Patella   (i) Sternum   (j) Pectoral girdle

10(a) Neural spine   (b) Neural arch   (c) Centrum   (d) Transverse process

Self test 4

1(a)   **A** Horse forelimb   **B** Bat forelimb

(b)		A	B
	Humerus	**A** Shortened and thickened	**B** Shortened and thinner
	Radius	**A** Fused, shortened and thickened	**B** Elongated
	Ulna		**B** Thin and reduced, fused at base to radius
	Carpals	**A** Elongation and thickening of metacarpal 3, 2 and 4 reduced and fused to 3, 1 and 5 absent	**B** Reduced and fused
	Phalanges	**A** Thickening of 3, nail becomes hoof	**B** 2–5 greatly elongated, 1 somewhat elongated

2(a)
Sutural	– fibrous connective tissue
Cartilaginous	– cartilage disc, connective tissue, ligaments
Synovial	– joint capsule of ligaments, synovial membrane, synovial fluid in synovial cavity, articular cartilage

(b)
Sutural	– little or no movement
Cartilaginous	– limited movement
Synovial	– free movement (exact movements depend on type of synovial joint, such as hinge, ball-and-socket)

3(a) Synovial – gliding
(b) Synovial – ball-and-socket
(c) Synovial – hinge
(d) Cartilaginous
(e) Synovial – saddle

4(*a*) Long bones act as the basic structure of the limbs, and as levers.

(*b*) Muscles bring about movement of the limb by contraction.

(*c*) Joints enable bones to move relative to each other.

(*d*) Nerves control the movement.

5 Proprioceptors are sensory nerve endings distributed in muscles and tendons. They play a role in keeping postural muscles at the correct degree of contraction or tonus. Because they remain sensitive even to a constant level of stimulation they are able to supply the brain with information about the state of each muscle.

6 The contraction of one muscle of an antagonistic pair will bring about the relaxation (stretching) of the other muscles. Figure 73 shows the muscles involved in bending and straightening the leg. They are the quadriceps femoris which straightens the leg and the biceps femoris which bends it.

7 See figure 178.

178   Sketch diagrams of kymograph traces: (a) single stimulus; (b) 8 per s stimulus; (c) 25 per s stimulus

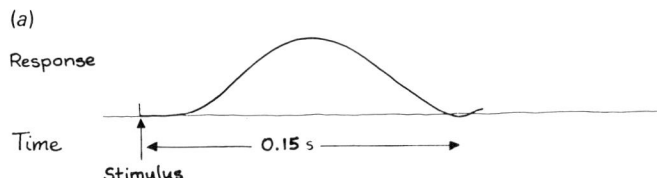

(*a*)

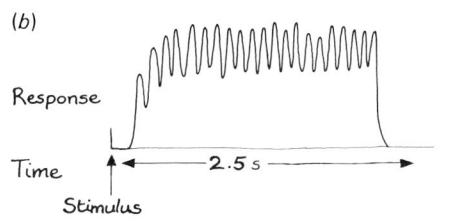

(*b*)

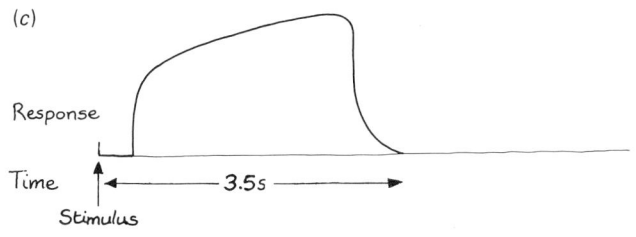

(*c*)

8 *Twitch*: This is a quick, sharp contraction produced as a result of a single shock.

Summation: When two or more stimuli are presented close together the responses combine together. For example a stimulus applied just after maximum contraction may cause the muscle to contract still further; the two separate contractions have been 'added' together or summated.

Tetanus: When a series of stimuli are applied above a certain frequency, a state of maintained contraction results known as tetanus.

Fatigue: After repeated and extended stimulation of a muscle, the response gradually decreases and disappears. This may be due to several factors including depletion in supplies of acetylcholine at the motor end-plate or accumulation of lactic acid. This is known as fatigue.

An *all-or-nothing response* occurs only after a stimulus has attained a certain threshold. When the stimulus increases beyond the threshold there is no increase in the response. The *refractory period* is the time after a response during which contraction does not usually occur. It is divided into an absolute refractory period in which no contraction can occur as the nerve axon supplying the muscle is inexcitable, and the relative refractory period in which only a strong stimulus produces contraction. The refractory period represents the movement of ions back to their original positions to re-establish the resting potential.

9(*a*) See figure 179.

(*b*) Class 3 lever

(*c*) Low mechanical advantage

(*d*) This means considerable effort must be expended to move the load (the arm).

179   Muscle involved in raising the arm

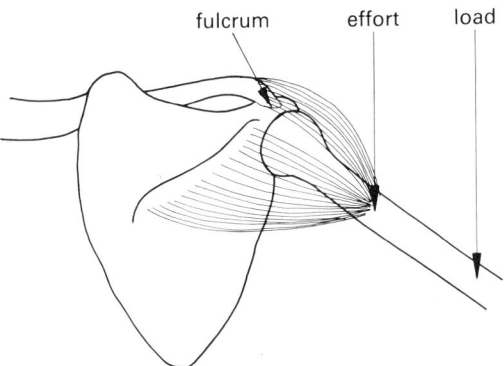

10 Tendons attach muscles to bones. They may be quite long. This allows muscles to act at a considerable distance from the bones they move. This is important, for example in the fingers where the bulky muscles closely associated with the finger bones would restrict the manipulations possible by the fingers.

11(*a*) The cerebellum receives input from the skin, muscles, tendons, the eyes and the canals of the inner ear and with this information it controls the balance and precision of movements.

(*b*) The cerebral hemispheres are concerned with whole sequences of movement, the learning involved in movement and the directedness of movement as a coordinated whole.

12 See figure 180.

180 The source and direction of forces which may act on a tree

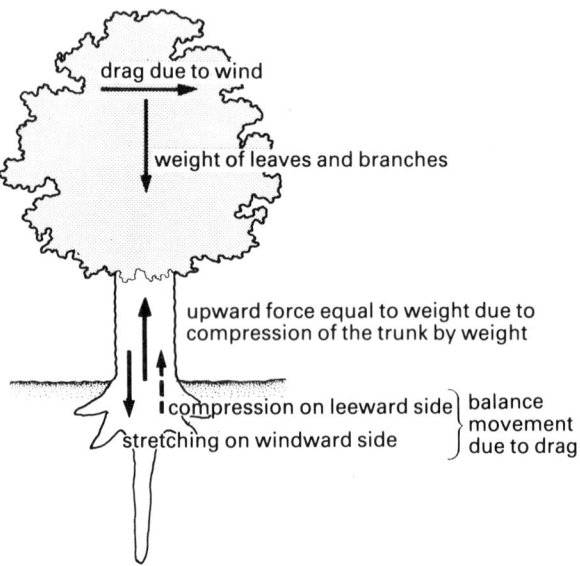

Self test 5

1(*a*) Amoeboid movement is carried out by certain protozoa and some other cells, such as vertebrate white blood cells. It involves the extension of projections of the cytoplasm known as pseudopodia. The rest of the body then flows forward into the pseudopodium.

(*b*) At the advancing tip of the cell, the central more liquid sol portion of the cytoplasm changes into the more jelly-like gel cytoplasm found around the periphery of the cell. At the posterior end of the cell, the corresponding change of gel to sol occurs.

2 See figure 181.

181 Comparison of movement in *Paramecium* and *Euglena*

	Paramecium	*Euglena*
Locomotory structure	cilia	flagella
Type of movement	ciliary movement	flagellar movement and contractile movement of cytoplasm
Direction of movement	can swim backwards as well as forwards	can only swim forwards
Strokes	downstroke of cilia is strong and straight, recovery stroke is limp	helical waves of movement pass along the flagellum
Movement of body	body rotates as moves forward	body rotates as moves forward
Speed	faster than *Euglena*	slower than *Paramecium*

3 The *body wall* of the earthworm contains layers of longitudinal and circular muscle. Alternate contraction of these layers causes the body to become short and fat or long and thin. Waves of contraction passing through these muscle layers allows the body to stretch forward, followed by a drawing up of the posterior end. In this way the earthworm progresses forwards.

The *coelom* is a fluid-filled space which acts as a hydrostatic skeleton against which the muscles of the body wall can act.

The *chaetae* can be extended and grip the substrate, thus anchoring the anterior end of the body while the tail is being drawn up. While the anterior end is stretching forward, the chaetae are withdrawn into sacs in the body.

4 A bends the knee (flexion).
B pulls the leg backward (retraction).
C pulls the leg forward (protraction).
D straightens the leg at the knee (extension).

5 *Myotomes* are blocks of muscle arranged along the body of a fish. A wave of contraction passes down the body on one side followed, after a short delay, by a similar wave on the other side followed by another wave on the first side, and so on. These contractions move the body from side to side. This causes the body and tail to push back and sideways against the water, so propelling the fish forwards.

The *tail* presents a large surface area against the water, so when it is moved from side to side by the myotome action it pushes back and sideways against the water, so propelling the fish forward. In teleosts, the tail is symmetrical and has no effect on altering the position of the fish in the water. In elasmobranchs, the tail is asymmetrical, the ventral portion being more developed than the dorsal portion. Its side-to-side movement, therefore, helps produce a lift to raise the fish up in the water.

The *vertical* and *horizontal fins* help maintain stability by preventing rolling.

The *horizontal fins* help prevent pitching. Pectoral fins are anterior to the centre of mass and so tend to make the fish unstable in pitch. The horizontal pectoral fins of cartilaginous fish help to produce lift to raise the fish up in the water.

The *swim bladder* of teleosts is used to maintain the height/depth of the fish in the water. Fish change their depth by swimming and subsequently adjust the amount of gas in the swim bladder. This enables them to remain at the chosen depth.

6 See figure 182.

182 Main forces acting on a bird's wing during flight

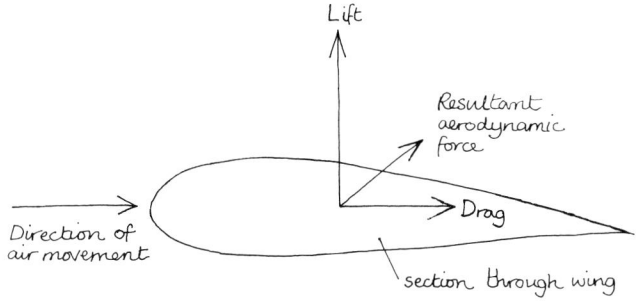

7(*a*) Lift is affected by the shape and size of wings and the exact angle of insertion of the wings.
(*b*) Speed of gliding is affected by the weight of the bird and size of the wings.

8(*a*) The downstroke is mainly associated with lift.
(*b*) The upstroke is mainly associated with forward propulsion.

9 Adaptations for flight include:
(i) Feathers are lightweight so less energy needs to be expended to keep the body in the air.
(ii) Skeleton is lightweight due to air-filled bones (advantage as (i)).
(iii) Limbs modified as wings to provide flight movement.
(iv) Feathers provide large surface area for providing thrust against air.
(v) Very powerful pectoral muscles to bring about wing movements.
(vi) Large keel for attachment of strong flight muscles.
(vii) Rigid skeleton provides stable framework for muscle attachment.
(viii) Fusion of bones creates rigid skeleton and reduces overall weight.

10 Insects fly by means of flattened wings attached to the thorax. The wings act as levers. They are moved up and down by changes in the shape of the thorax caused by contraction of two main sets of muscles acting on a series of cuticular levers (figure 183).

Contraction of the vertical muscles flattens the thorax dorsoventrally and causes the wings to move up. Contraction of the longitudinal muscles shortens the thorax anterio-posteriorly and causes the wings to move down.

183 Muscles associated with flight in insects

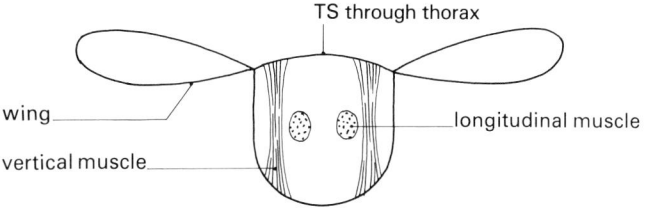

Self test 6

1(*a*) Lorenz is sometimes called the 'father of modern ethology' because he is generally acknowledged as its founder who laid the groundwork for many lines of research still being followed today.
(*b*) He believed in a natural approach, studying animals in their own surroundings. Most of his findings came from observation rather than experimentation.
(*c*) The phenomenon of imprinting.

2 *Photography* of various kinds, such as stills, cinefilms, time-lapse and multiple-flash, all allow a more prolonged and detailed analysis to be made of visual sequences than can be obtained with the unaided human eye. *Cinefilm and multiple-flash* are particularly useful for the detailed study of movements. *Time-lapse* enables very slow changes to be detected as complete movements. *Stills* allow every detail of a scene to be detected.
Tape-recording allows for detailed analysis of sounds, especially song patterns, and tapes may be played at slow speeds for this purpose.
Video-filming is useful for studying behavioural sequences.
Activity recorders detect longer-term patterns of activity.
Microprocessors store information for later detailed analysis.

3 Models are particularly useful when investigating 'sign stimuli' or important visual signals which release certain kinds of stereotyped response. Possible significant features can be presented separately and the animal's response measured. Thus, shape, colour or positioning of features may be investigated separately and in combination and their contribution to the behaviour assessed.

4 *Fixed action pattern*: A typical pattern of motor actions shown by all members of a given species. A form of stereotyped behaviour, such as the egg-rolling response of the greylag goose.
Sign stimulus or releaser: The exact stimulus that initiates behaviour may often be broken down to something quite small and specific. In territorial disputes between robins, threat behaviour is shown in response to a tuft of red feathers almost as readily as a stuffed adult bird. A stuffed brown bird produces little reaction. Here, it appears that red colour is the sign stimulus.
Supernormal stimulus: A stimulus much larger than that which is normal, will attract a response, often a stronger one, than the natural stimulus presented at the same time. For example, ground-nesting birds may retrieve a large model egg in preference to their own egg. A warbler may respond to the large 'gape' of a baby cuckoo and feed it in preference to its own young with smaller 'gape'.
Reflexes: The simplest, unlearnt responses found in organisms possessing a nervous system whose nature is determined by an inherited pattern of receptors, nerves and effectors, such as the knee-jerk reflex or 'startle' reflex.

5(*a*) (i) Innate behaviour is stereotyped for the species, learned behaviour is not stereotyped.
(ii) Innate behaviour is not affected by experience, learned behaviour is affected by experience.
(iii) Innate behaviour is not adaptable or flexible, learned

behaviour is adaptable or flexible, learned behaviour is adaptable.
(iv) Innate behaviour is genetically determined. The ability to develop learned behaviour patterns is inherited but not the actual behaviour patterns.
(*b*) Many examples of behaviour cannot be described as either innate or learned, which the use of such terms might suggest. This could lead to the danger of over-simplifying explanations of behaviour.

6(*a*) Insight learning/reasoning
(*b*) Chemotaxis
(*c*) Trial-and-error learning and latent learning
(*d*) Reflex action and habituation
(*e*) Operant/instrumental conditioning
(*f*) Orthokinesis

7 Both taxes and kinesis are concerned with the *orientation* of an organism to some aspect of its environment. When the stimulus does not control the *direction* of movement, the response is a kinesis. Taxes always show a relationship between the direction of movement and the direction of the stimulus.

8 *Imprinting*: When a young bird or mammal is hatched/born, it will follow around the first moving object that it sees (usually its mother). This is known as imprinting.

Imprinting only occurs during one critical time period towards the beginning of an animal's life whereas other kinds of learning may occur at any age. Imprinting determines subsequent behaviour in an inflexible way. This is not a characteristic of normal learning.

9 Operant conditioning 'shapes' the behaviour of an animal by the process of reinforcement (rewarding). A form of trial-and-error learning is involved. This procedure differs from classical conditioning in which the stimulus for a reflex behaviour is paired with an irrelevant stimulus which ultimately replaces it. Learning is involved in the 'replacement'. Thus, operant conditioning results in learned responses to a given stimulus while classical conditioning involves basically an unlearned response to a stimulus that would not previously have elicited the response.

10 A grasp of a concept or the understanding of a simple set of relationships may be described as a learning set (such as the understanding of the idea of 'oddity'). Such a 'set' may be achieved by trial-and-error learning but is then transferable as a unit to a new situation so that the individual does not always 'start from scratch' when faced with a new problem.

Self test 7

1 The hypothalamus appears to be the seat of certain basic 'drives', such as thirst, sleep, sexual and maternal behaviour. These 'drives' strongly motivate an animal to attain the goal of the behaviour. A thirsty animal will seek to drink and reduce the need it feels and its behaviour will be directed towards achieving that end.

2 The changing length of the photoperiod in temperate regions must be detected by some receptor inside an animal, possibly the pineal gland which, in response, triggers the activation or renewed production of the sex hormones such as testosterone and oestrogen. These hormones influence the behaviour of the animal making them actively seek out or be receptive to potential mates, set up and defend territories prior to courtship, and so on.

3 *Annual rhythms*, such as the reproductive rhythms of sheep or hibernation of tortoises.
Daily rhythms (circadian rhythms), for example alternation of sleep and activity.
Tidal rhythms, for example some marine snails living high up the shore release their reproductive cells at the time of spring tides only.

4 The worker bee is communicating information about the distance and direction of a food source. The speed of the dance is related to the distance of the food source from the hive. The direction of the food source is communicated by the orientation of the bee's body.

5 A pheromone is a chemical substance produced by an animal which influences the behaviour or development of an animal of the same species. Ants produce chemical trails that guide other worker ants to a food source. Female moths produce chemicals which attract male moths for mating purposes. The odour of a male mouse may initiate the oestrous cycle of a female. Many male mammals secrete chemicals which serve as territorial markers.

6 Species recognition for reproductive purposes, territorial defence, alarm or warning

7 Courtship serves to bring the sexes together: it may suppress aggressive behaviour, synchronise reproductive behaviour for maximum effectiveness and ensure that breeding takes place only between members of the same species.

8 It offers protection from predators, may increase feeding efficiency, may increase reproductive efficiency

9 *Agapornis* sp. show an evolutionary trend from the building of simple, unshaped nests to those of elaborate nests entered through a tunnel. A trend is also revealed in the way the nesting materials are carried, the ancestral species carrying nesting materials tucked into their feathers and later species carrying in their bills only.

The hybrid offspring of the peach-faced and Fischer's lovebirds show a *conflict* in the way they carry nesting materials, trying both methods on occasion but eventually, over the years, carrying nesting materials nearly always in their beaks. This could be due to their inheritance from the peach-faced lovebirds which mostly carry materials in their feathers and the Fischer's lovebird which is a beak-carrier only. This throws light on genetic control of behaviour.

10(*a*) *Drosophila* with the gene for bar eye expressed in the phenotype will see less well than normal *Drosophila* and the visual aspects of courtship may not be so well detected by them or responded to.
(*b*) Wing vibration is an important part of the *Drosophila* mate's courtship and males with dumpy wings may perform less effectively than those with normal wings.

This information shows that behavioural performance can be affected by genetic inheritance and shows the effect of single mutations which are the raw material of evolutionary change.

Section 8: Answers to self-assessment questions

1 (*a*) Chewing; mechanical breakdown of food
(*b*) Running; avoid harmful situations
(*c*) Collecting; gathering useful objects (food, fuel, weapons, and so on)
(*d*) Facial expression; communication
(*e*) Routines; aid survival

2 (*a*) Gaining optimum illumination for photosynthesis.
(*b*) Dispersal of spores for successful reproduction.
(*c*) Gaining optimum illumination for photosynthesis.
(*d*) Pollination by night-flying insects.

3 Animals are unable to synthesise their own food and so need to be able to move to find and capture their food.

4 Aquatic – water movements (either natural or self-produced) can carry food to the animal (freshwater and marine).
Parasites of alimentary canal – food is collected by the host organism and brought to the parasite or attached to shell of another aquatic.
Intertidal or animal – for example anemones attached to hermit crab shells feed from debris of crab, goose barnacles on whales, and so on.

5 A tropism is a *directional* movement of part of a plant in response to a *directional* stimulus.

6 Active transport requires the expenditure of energy from the ATP → ADP + P reaction.

7 Photonasty

8 Nyctinastic movements; *Oxalis* plants

9 (*a*) The food it consumes
(*b*) Respiration

10 The nervous system

11 Overcome friction and resistance of medium.
Energy expenditure required.
Transform potential to kinetic energy.
Moving parts associated with contractile tissue.
Movement by backward and downward force.
Control mechanism to coordinate activities.

12 Protection of delicate organs, firm anchor for muscles, levers for movement

13 (*a*) The body will extend to the right and become thinner on the left side.
(*b*) The internal fluid pressure will increase.

14 If the animal keeps a constant length, the fluid pressure will cause the muscles on the right to be relaxed or even stretched as the body wall bulges outwards.

15 A set of longitudinal muscles

16 Mesenteries and fluid

17 The gullet will close and enclose the fluid in the enteron.

18 (*a*) Musculo-epithelial cells
(*b*) Longitudinal muscle band and circular muscle fibres in ectoderm

19 Oral sphincter muscles contract enclosing fluid in enteron which forms the hydrostatic skeleton. Longitudinal muscle band contracts, with more contraction on the right-hand side, causing the animal to bend in this direction. Circular muscles are relaxed.

20 Bone, cartilage

21 Sclerotin; insects or arachnids

22 Cellulose

23 Blood has the connective tissue characteristic that the cells are contained in a non-living matrix (the plasma).

24 (*a*) Yellow and white fibres
(*b*) Fibrocytes, mast cells and macrophages

25 The elastic property of the artery walls enables them to accumulate blood during systole and to discharge it by elastic recoil during diastole. Thus, the intermittent movement of blood out of the heart is converted to a steadier flow through the peripheral circulation.

26 Dense fibrous tissue differs from areolar tissue in that: fibrocytes occur in columns, not scattered through the matrix;
fibres run parallel with no branching;
no mast cells or macrophages/white or yellow fibres;
fibres are of collagen and not elastin.

27 Tendons must resist stretching so that when the muscle connected to it contracts, it transmits the force of contraction directly to the bone causing it to move.

28 Yellow elastic fibres

29 Shark, ray, dogfish or skate

30 The collagen fibres help produce a tissue of great strength and rigidity which is necessary as the vertebral column bears much of the weight of the animal. The collagen fibres also confer a certain amount of elasticity which means that the discs can be deformed during movement and regain their original shape.

31 Such cartilage is stretched and will then return to its original shape, that is, it is flexible.

32(*a*) Cartilage and bone
(*b*) Bone
(*c*) Collagen fibres
(*d*) Elastic fibres

33 Osteoblasts deposit new bone matrix while osteoclasts break down the calcified cartilage.

34 First in the central region of the shaft and later at either end.

35 *Cartilage bone*: (*a*) cartilage, (*b*) bone, (*c*) limb bones, vertebrate girdles
Membrane bone: (*a*) connective tissue in embryo, (*b*) bone, (*c*) skull bone, clavicle

36 See figure 184.

37(*a*) Actin and myosin
(*b*) Actin only
(*c*) Myosin only

38(*a*) Length unchanged
(*b*) Length reduced
(*c*) Shortens
(*d*) Remain the same

39 Roughly cylindrical. Length is greater than width and breadth. In TS they are polygonal – usually six-sided. The upper and lower ends of the cells may be tapering. Irregularly thickened cellulose wall especially at cell angles.

40 See figure 185.

185 Comparison of collenchyma and sclerenchyma

	Collenchyma	*Sclerenchyma*
Function	support	support
Cell wall	irregularly thickened with cellulose only	evenly thickened with cellulose and lignin
Shape	elongated longitudinally, roughly isodiametric (all diameters approx. equal in length) in TS	elongated longitudinally, roughly isodiametric in TS
Position	below epidermis	associated with vascular bundles and cortex
State at maturity	living cytoplasm in cell	no living contents

41 The xylem is very poorly developed.

42 In water, the fronds are separate and well spread out. Out of water they mat together.

43 The supporting and separating effect of the water on the fronds ensures adequate light absorption for photosynthesis.

184 Comparison of muscle cells

	Striated	*Smooth*	*Cardiac*
Size	1–100 mm × 100 μm	up to 0.5 mm × 6 μm	4–5 μm × 2–3 μm
Shape	elongated	elongated	short cylindrical
Structure	multinucleate and striated	single central nucleus no striations	striated columns of cells from fibre

44 Problems include lack of protection, lack of support for movement and increased chance of dehydration.

45 The mass will increase by the power of 3 but its exoskeleton only by the power of 2, that is relatively much less, though the exoskeleton has to provide its support.

46(*a*) Tension forces
(*b*) Compression, tension, shearing and torsion
(*c*) Compression, tension, shearing and torsion

47 The lizard has bow legs which require greater energy to keep them in a position to support the body off the ground than a mouse which has its legs positioned directly below the mass of the body.

The greater energy needed by the lizard is provided by the relatively larger adductor muscles. Larger muscles in turn require a stronger attachment surface so the pelvic girdle of the lizard is also relatively larger than that of the mouse.

48 Within the solid ground substance there are numerous cavities (lacunae) and tubes (canaliculi and the Haversian systems).

49 At the edges

50 (i) There is very little stress in the centre of a long bone. Therefore, it would be uneconomical to 'waste' bone tissue in this region.
(ii) It makes animals lighter. Therefore, they need less energy to move around.

Both scaffolding and bicycle frames are made of hollow tubes for the combination of strength with lightness.

51 Iliotibial tract

52 Being composed of dense fibrous connective tissue, it has great tensile strength.

53 The muscle can alter the tension of the tract and can therefore operate efficiently over a range of stresses.

54 There are two pairs of cantilevers (four girdles), each pair centred around the fore- and hindlimb girdles: that is the neck region cantilever (cervical) and thoracic region cantilever, both supported from the pectoral girdle, and the abdominal (lumbar) cantilever and tail (caudal) cantilever which are supported from the pelvic girdle. (The tail cantilever plays a reduced supporting role in the animal shown in figure 57.)

55 Both arm and leg conform very closely to the pentadactyl plan. The only differences are that, in the arm, the 4th and 5th distal carpals are fused and one proximal carpal is missing, and, in the leg, the 4th and 5th distal tarsals are fused and one proximal tarsal is missing.

56 See figure 186.

57 A Carries oxygen and glucose to provide for energy release during respiration so that muscles may contract and do work.
B Synovial joints enable bones to move smoothly relative to each other.
C Contraction of muscles attached to bones brings about movement of the limb.
D Nerve impulses from the central nervous system control and coordinate limb movement.
E Ligaments hold the bones together at joints. They help restrict movement of the bones and prevent dislocation of the joint.

58 Collagen fibres would predominate. They combine a high tensile strength with flexibility. Elastic fibres would be unsuitable as they would stretch too much and would not produce a tight joint.

59 Cartilage resists compression forces which are exerted at these joints. It would also act as a shock absorber.

60(*a*) Hinge, (*b*) ball-and-socket, (*c*) pivot, (*d*) gliding, (*e*) saddle, (*f*) hinge and modified rotating gliding joint

61(*a*) Biceps femoris
(*b*) Contraction of the biceps femoris causes the leg to bend, by pulling the fibula via the internal hamstring tendon, and this movement of the bone pulls on the patellar tendon of the quadriceps femoris, thus stretching this muscle to its relaxed length. When the quadriceps femoris is stimulated to contract, it pulls on the tibia bringing it forward (extends it) and causes the biceps femoris to be stretched.

62(*a*) The potential difference across the membrane of a neuron that is not conducting a nerve impulse.
(*b*) The localised change of electrical potential across the membrane of a neuron associated with the passage of a nerve impulse.
(*c*) The time between the application of a stimulus and the first detectable response in the nerve.
(*d*) The strength of a nerve impulse is not varied by varying the strength of the stimulus. The impulse either occurs in its normal form or it does not.
(*e*) The removal of the polarisation of the neuron membrane during the passage of a nerve impulse. The resting membrane is polarised so that the inside is electrically negative to the outside.
(*f*) The additive effects of separate nerve impulses arriving at a particular neuron. Spatial or temporal summation can

	Whale	Horse	Bat	Mole	Rabbit
Humerus	much shortened and thickened	shortened and thickened	thin and elongated	shortened and thickened	shortened, sturdy, at right-angles to radius
Radius	shortened	fused, shortened and thickened	elongated	shortened and thickened	elongated, separate but tightly bound together
Ulna	shortened and thickened		thin and reduced, fused at base to radius	elbow enlarged, reduced and partly fused to radius	
Carpals	reduced in number	elongation and thickening of metacarpal 3, 2 and 4 reduced and fused to 3, 1 and 5 absent	reduced and fused	extra bone present; sickle-shaped radial sesamoid bone	8/9 present, reduction due to fusion, metacarpals elongated
Phalanges	2 and 3 lengthened, others reduced	thickening of 3, nail becomes hoof	2–5 greatly elongated, 1 elongated	other bones all broad, broad claws	1 short, 2–5 elongated, horny claws cover terminal phalanges
Function	swimming (paddle)	running	flying	digging and walking	burrowing and receiving 'shock' of alighting from jump

induce either stimulation or inhibition of nervous activity.

(g) The intensity of a stimulus below which there is no response by a neuron.

(h) The period of time after the conduction of an impulse during which the neuron cannot transmit another impulse.

63 A frog is an ectotherm, so maintaining a constant temperature for the nerve–muscle preparation is not of such great importance as it would be for a mammalian preparation.

64 See figure 187.

187   Labelled muscle response

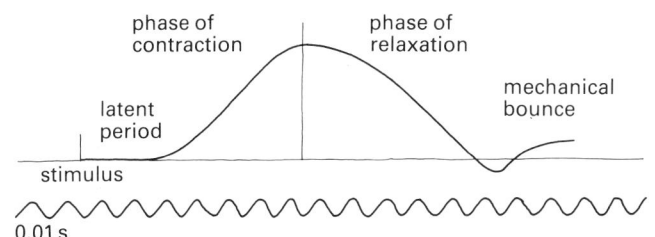

65 14×0.01 s $= 0.14$ s.

66 The gradual contraction results from the progressive increase in the number of motor units within the muscle contracting as the stimulus spreads across the muscle.

67 A single smooth contraction followed by relaxation.

68 *0.02 s interval:* The response is of the form shown in the simple muscle twitch. The second stimulus had no effect on the muscle as it was in process of contraction.
0.1 s interval: Two contraction responses are becoming evident although the second response commences before complete relaxation of the first contraction. The second contraction shows increasing amplitude over the first.
0.3 s interval: Here, two separate responses of the simple twitch form are produced. Again, the second contraction shows a slightly greater amplitude than the first.

Summation is shown at the 0.1 s interval only.

69 As the frequency of stimulation increases, the muscle goes into sustained contraction until the stimuli stop.

70 With muscles that move the fingers situated further back along the limb, the fingers themselves can be very much thinner and therefore capable of much finer movements and actions.

71 Further away

72 Decreased

73 The weight of the head bent forward is balanced by the action of the muscles in the back of the neck. As they are used such a lot, they will become fatigued.

74 They spend a lot of time dancing on the tips of their toes. To maintain this position, the calf muscles must be well developed to provide the effort to act against the load acting down through the body.

75 Muscle appears to develop in response to use. There is some sort of feedback between the mechanical output of the tissue and the process of growth and maintenance. The lifting of heavy loads requires much work from the flexor muscles of the arm which develop so that even greater force may be exerted.

76 Weight of branches and leaves and organisms associated with them

77 Drag

78 When the wind blows.

79 A colloid is a mixture of very small particles suspended in a continuous medium. The mixture has properties between those of a solution and fine suspension. Colloid may exist as sol (fluid) or gel (semi-solid) forms.

80 Endoplasm or plasmasol is relatively fluid and contains many granules and organelles. Ectoplasm or plasmagel is semi-solid and contains few particles.

81 Flexion, protraction, extension, retraction, flexion, protraction, extension...

82 Retraction

83 Flexion, **E**; protraction, **A**; extension, **B**; retraction, **F**

84 (*a*) **C** (*b*) **D**

85 Greater

86 The upper body inclines forwards and the arms are flexed to a greater extent.

87(*a*) Fulcrum – toes (phalanges); effort – tendon and muscle **C**; load – the weight of the body acts at the joint between tibia and calcaneum.
(*b*) Downwards and backwards thrust

88 Elongation of metacarpal 3 (also thickening)

89 Radius, ulna and metacarpals all elongated.

90(*a*) The whole foot – tarsals and phalanges
(*b*) The hoof (tip of phalange only)
(*c*) Phalanges
Man's foot does not contribute to the length of his leg except when he rises on his toes to run. The rabbit, to some extent, and the horse very markedly, have lengthened limbs and therefore the increased length of stride due to rising up on part or all of their foot. Length of stride is one factor influencing speed.

91 1–4 Tail moves to right of midline of body
5–7 Angle of tail to midline of body changes
8 Tail moves back to midline
9–10 Tail moves to left
11–12 Angle of tail changes
13–16 Tail moves to right again

92 See figure 188.

188 Muscle contraction during locomotion in a fish

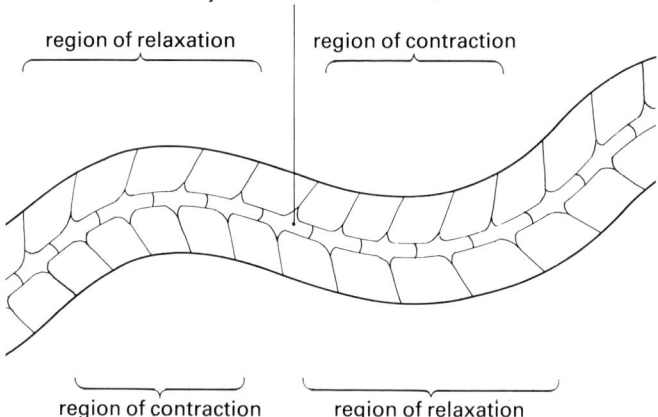

93(*a*) Horizontal pectoral and pelvic fins, but in teleosts the pectoral fins are anterior to the centre of mass and so tend to make the fish unstable in pitch.
(*b*) Vertical dorsal and ventral fins; also the paired horizontal fins.

94 It will tend to sink.

95(*a*) They may escape from non-flying predators and exploit different food reserves.
(*b*) Organs concerned with detecting body position to make them very responsive to small changes of angle and direction?

Touch receptors on body surface to make them aware of wind speed and direction?
Possibly increase of sensitivity in organs of vision which may need to be used over greater distance?

96 Shape and size of wing, exact angle of insertion of the wing, wind speed

97 As the mass of the bird decreases, the rate of wingbeats increases.

98(*a*) The keel is very large. It acts as the anchoring area for the large flight muscles and must be able to withstand the powerful forces which act against it during flight.
(*b*) Synsacrum (or parts of the limbs)

99(*a*) Levator
(*b*) Depressor

100 Lift is associated with the downstroke. It is achieved by the downthrust as illustrated in (2) and also by the wings acting as aerofoils. There is also some lift at (6) in the upstroke.

Forward propulsion is also generated by the powerful downstroke in the obliquely downward and backward movement of the wing.

101(*a*) In the crane-fly, the hindwings have been modified to form sense organs, called halteres.
(*b*) In the bee, the fore- and hindwings are linked together by bristles/hooks and function as a single wing.

102 Diptera: for example housefly, fruitfly, hoverfly, horsefly, mosquito (there are 5000 British species!)

103 A load; **B** fulcrum; **C** effort

104 Small movements of the 'effort' (thorax wall and muscle) produce large movements in the load-bearing wings.

105 Insect flight muscles possess tracheoles that pass between the fibres (as do the capillaries in vertebrate muscle). The oxygen simply diffuses inward from the surface of the fibre. Some tracheoles pass deep into the fibres without actually penetrating them (like fingers pushed into a balloon). This decreases the diffusion distance.

106 Fabre has constructed an experimental situation to test a particular hypothesis:
– he observes constantly, with attention to detail (and repeats observations over a period of time);
– he clearly keeps a careful record of all his observations;
– his inferences or conclusions depend solely on what he has observed and do not go beyond the evidence of observation.

107 The beetles are attracted to dead animals by smell. They crawl under the body and the weight on their backs stimulates digging movements. If the substrate is unsuitable, the beetle will turn onto its back and grip the skin or fur with all its legs and try pushing the dead animal. With no communication or cooperation, a number of beetles may move the food to a suitable substrate. Once a large enough hole is dug, female beetles will lay eggs onto the dead animal.

108 The male Muscovy tried to mate with Greylag geese; males or females alike.

109 Both examples involve leaving the birds in as natural an environment as possible while manipulating one specific aspect of the environment (isolation from own kind).

110 Imprinting occurs early in life: the 'following' reaction for the Greylag was established within one week. The Muscovies only remained with their foster parents for seven weeks. Imprinting is irreversible and will remain fixed on the initial subject. Imprinting may affect sexual behaviour in later life.

111 Stills and cinefilms allow a more prolonged and detailed analysis of passing events. Time-lapse photography allows longer periods of time to be observed by sampling at regular time intervals, allowing the observation of slow or otherwise imperceptible movement. Multiple-flash can show events normally too fast for single exposure photography and show the changing relationship of, for example, parts of the body in one picture.

Video-filming is useful for recording complete behavioural sequences.

112 Most animals cannot detect infra-red light and so the trip beam does not affect the behaviour to be studied.

113 The most obvious sex difference seen in figure 132 is the dark patch beneath the eye of the male. Exposing a male to a variety of models with and without the 'male patch' and other markings may show which sign is important.

To test the conclusions, a female bird can be marked to resemble a male in the crucial factor and then observe the male response.

114 An unnatural food source (sugar solution with lavender oil) is provided. Scent plates are set up in various places.

115 Bees have good vision. If passing feeding bees on their flight they may have been attracted to stop there and feed; also some bees might return to the hive and give different 'distance' information from the 'new' food sources.

116 The stimulus for a reflex behaviour is paired with an irrelevant stimulus which is ultimately able to stimulate the given reflex by itself.

117 Red coloration of breast feathers

118 Red dot on yellow bill (level of contrast rather than colour being important). Thin and elongate bill near to chick, low, and with the tip pointing downwards. Hunger of the chick is also necessary.

119 The movement is stereotyped and does not adapt to changing circumstance: it continues to completion, even if the egg has rolled away in the course of the retrieval. When the egg is again seen outside the nest, the movement will then be repeated.

120 The large gape of a cuckoo chick must be a supernormal stimulus.

121(*a*) Insects
(*b*) Human beings
(*c*) Simple metazoa

122 Klinokinesis

123 No

124 In the dark, *Dendrocoelum* changes direction at approximately $100°$ min $^{-1}$. This rate of change of direction increases eight times when a light is switched on and returns over roughly 30 min to its previous level.

125 The planarians become adapted to the new higher light intensity level.

126 Dry surroundings stimulate woodlice to move about. They stop moving in damp conditions.

127 Animals are kept at a constant humidity. There is no directional component to the stimulus. Possibly orthokinesis where speed of movement is concerned. There is no mention of turning which might indicate klinokinesis. However, the evidence is not clear enough for firm conclusions.

128 It maintains the woodlice in favourably damp environments which would protect them against dehydration.

129 Klinotaxis

130 Operant conditioning actually 'shapes' the behaviour of the animal by the process of reinforcement, and trial-and-error learning is involved. Classical conditioning

does not alter the actual behaviour of the animal and learning is only involved in the replacement of a natural stimulus by an irrelevant one, through association of the two stimuli.

131 15 days

132 Experience of mazes allows rodents to negotiate them with fewer errors than previous trials (trial-and-error learning).

133 Group **1** show a continual steady decline in mean number of errors over 15 days. Group **2** shows a slightly lower reduction of error for 6 or 7 days followed by a levelling of performance with no further improvement.

134 The provision of a food reward at the end of the maze provided a reinforcement for learning the most direct route to the food (that is increased motivation to learn).

135 It ensures the young remain close to the mother who provides them with food and shelter and protects them from predators.

136 Very early in life – maybe within the first week. No, an animal imprints on one object only.

137 It tried mating with male or female Greylag geese rather than females of its own species.

138 Here the 'reaction' is the mating behaviour and the 'object' of the reaction is normally females of the same species. In the Muscovy duck, the mating 'object' is laid down very early on (male or female Greylags), well before the maturing of the sexual reaction.

139 16 hours in the mallard

140(*a*) The further a mallard duckling is allowed to follow an object during the critical period, the more strongly imprinted on that object it will be. (Strength of imprinting judged on likelihood of subsequently following the given object.) Up to 50 ft distance imprinting grows in effectiveness but then levels off.
(*b*) Effort appears to be related to the imprinting process.

141 Practice improves the accuracy of pecking, a two-day old chick performing as well as a four-day old chick that has not had practice.

142 Internal receptors detect that blood sugar is low or that the stomach is empty (hunger drive). This causes the baby to wake (hunger is stronger than the need for sleep). The baby cries because that is its way of attracting food (learning may strengthen this behaviour). The baby suckles (feeding behaviour) until it is no longer hungry when it will stop. The hunger drive is reduced and the need for sleep is again stronger.

143 Adrenaline – 'fight or flight'.

144 Gonadotropin (FSH), oestrogen, progesterone and prolactin

145 Courtship

146 Fertilisation of eggs and progesterone levels of the female

147 Onset and maintenance of incubation triggers prolactin production and crop milk.

148 The rate of testicular development reaches a maximum for a number of species when day-length reaches 16–18 h. The lack of sexual maturity during the shorter winter days of low temperatures and scarcity of food does not allow reproduction in these less-favourable conditions.

149 Lambs tend to be born towards the end of winter as the days are lengthening (such as March). Mating would then occur around October when the photoperiod is *decreasing*.

150 Distance, direction and scent of food source

151 Communication was by a dance and the scent on the scout bees.

152 The angle of orientation of the dance with respect to the vertical indicates the direction of the feeding place with respect to the Sun: vertically up, food on a line between hive and position of Sun; vertically down, food on a line with hive and Sun, but in opposite direction to Sun.

153 Prolactin stimulates milk production after giving birth and may be responsible for maternal behaviour. However, it also acts synergistically with LH to maintain the corpus luteum and hence the progesterone it secretes which is essential in maintaining pregnancy.

154 Ensures that the eggs are laid in a specific and protected area – the nest.

155 The act of a male fanning a nest may indicate to a female that eggs are present in that nest and make her more likely to enter into the courtship sequence. (Thus, the male 'deceives' the female.)

156(*a*) *Agapornis* spp. show an evolutionary trend from carrying nesting materials tucked into body feathers in Madagascar lovebird, through partly tucking material into lower back feathers and supplementing this with beak-carrying (peach-faced lovebird), to carrying material

piece by piece in beak only (Fischer's lovebird).
(*b*) The behaviour of the hybrid (peach-faced and Fischer's) is intermediate between that of the parent species, the conflict being finally resolved by beak-carrying.

157 Instinctive behaviour is genetically determined as shown by the two methods of carrying nesting materials originally displayed by the hybrid. However, the hybrids learn by experience that beak-carrying is more effective and gradually adopt that pattern in the main. Their later occasional attempts to tuck materials into tail feathers are better performed showing improvement brought about by practice on an originally unlearned behaviour. This shows the close interrelation in practice of the two behavioural types.

158 The nest of the Madagascar lovebird is simple and exposed and its eggs consequently more vulnerable to predators than that of the other lovebirds with their more elaborate nest which needs less defence.

159 (i) Yellow males have a lower vibration percentage than wild-type males.
(ii) Yellow males have a shorter bout length than do wild-type males.
(iii) The orientation phase of yellow males is longer than wild-type males.

160 The abnormal wings cannot vibrate in the way necessary to properly stimulate the female's sense organs.

Pre-test: Levers

1 A lever is a rigid bar which may be turned freely about a fixed point.

The fulcrum is the fixed point of support about which the lever can turn.

The effort is the force applied to the load/resistance to bring about movement.

The resistance is the force exerted by the structure which will be moved about the fulcrum.

Mechanical advantage is a measure of how much effort is required in relation to the load to be moved. The greater the mechanical advantage, the smaller the effort required in relation to the load.

2	**A**	**B**	**C**
Pliers	load	fulcrum	effort
Bottle opener	effort	load	fulcrum
Sugar tongs	load	effort	fulcrum

3(*a*) First class lever
(*b*) Second class lever
(*c*) Third class lever
See figure 189.

189 Classes of lever

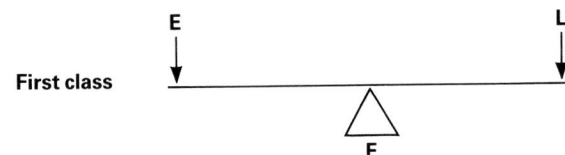

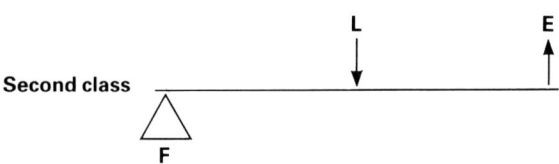

4 The function of a first class lever is to lift a heavy load with a minimum effort.

The function of a second class lever is to lift a heavy load with a minimum effort.

The function of a third class lever is to cause a large movement of the load for a small movement of the effort.

Index

Numbers in italics signifies figure reference.

acetylcholine (ACh) 19, 44
Achilles tendon (*tendo calcaneus*)
    44
actin, 17, 51
activity recorders 75
activity wheel *75*
adenosine triphosphate (ATP)
    3, 18, 19
adipose tissue 10
aerenchyma 23
air, movement in 63
American woodpecker,
    male/female 77, *77*
Amoeba 51–3
    movement 52–3, *52*
aquatic organisms 3
arches 34
arm, human, bones 36
arthrodial membrane 24
arthropods
    antagonistic muscles in limb
        59
    exoskeleton (cuticle) 23–4,
        24
    walking 59
ATP *see* adenosine
    triphosphate
avoiding reaction 86

Barbary dove *see* ring dove
behaviour 70–109
    analysis 76–8
    classification 83
    evolution 104
    importance in mammals 71
    inner readiness 83
    instinctive (innate) 83
    observing and recording
        81–3
    physiological determinants
        83
    sign stimuli (releasers) 83
    social 104
    stereotyped (fixed action
        patterns) 83, 84

trends in animal evolution
    85
biological rhythms
    (endogenous rhythms)
    98–9
birds
    flight 63–4; Canada goose *65*
    flight apparatus *64*
    forces on wing during
        gliding *63*
    migration 100
    rates of wing beat 64
    skeleton 64, *64*
    song 102
    wing structure *65*
blackbird chattering 100
bladderwort (*Utricularia*) 4, *4*
bone 11–13
    canaliculi 11
    cartilage 12
    compact *11*, 12, 30
    Haversian system 12, *12*
    lacunae 11
    long, growth and
        development 12, *12*
    membrane 12, *13*
    recycling 12
    red marrow 30
    resistance of environmental
        forces 30–1
    skull 13
    spongy 11, *11*, 30
    trabeculae 31, *31*
braces 31–2, *32*
bridges, model *34*
brine shrimp (*Artemia*),
    orientation 90

Canada goose, flight
    movements *65*
cantilevers 32–3, *33*
cartilage 10–11
    elastic 11
    hyaline *10*

centre of gravity of body 27–8,
    28
cerebellum 44
cerebral hemispheres 44
chemotropisms 3
chicks, pecking accuracy 97
chimpanzee, insight learning
    93
chitin 9, 24
chondrocyte 10
cilia 53, *53*
circadian rhythms 99
classical conditioning
    (conditioned response,
    reflex) 78–9, 90
cleaner fish 100, *100*
collagen 8, *8*
    fibres 9
collenchyma 21, *21*
communication 100–4
conditioned response *see*
    classical conditioning
conditioning 90
    instrumental (operant) 90–1
connective tissue 9–10, 13
    areolar 9, *9*
    fibrous 10,13
    matrix 10
counterbalancing 32, *32*
courtship 102–3
cross-species studies 104–6
cuticle 23–4, *24*
    advantages/disadvantages
        25
    moulting (ecdysis) 24
    muscle attachment 24

Darwin, Charles 72
Dendrocoelum, kinesis 86, *86*
detour problem 93
Difflugia 5
dogfish, locomotion muscles *61*
drives 97
Drosophila melanogaster (fruit
    fly) 106–7

earthworm 54, *54*
    coelom 54
    locomotion 55, *55*
    segments 54
ecdysis (moulting) 24
elastic fibre 9–10
endocuticle 24
endogenous rhythms *see*
    biological rhythms
endoskeleton 5
epicuticle 24
ethogram 81
ethology 70
Euglena 54
evolution of intelligence 96
evolution of vertebrate limbs
    29
exocuticle 24
exoskeleton 5
    advantages/disadvantages
        25
    arthropods 23

fatigue 44
femur, human
    stresses on 31, *31*, *49*
    trabeculae 31, *31*
fibre 7
fibroblast 10
fish
    buoyancy 61–2, *62*
    cleaner 100, *100*
    pectoral fins 62
    pelvic fins 62
    stability/instability 61, *61*
    swimming movements *60*
    swimming muscles 61
    tail fins 62, *62*
fixed action patterns
    (stereotyped behaviour)
    83, 84
flagella 53
flatworms, orientation 87–8
flour beetle (*Tenebrio molitor*)
    89

fly larva, response to light 88–9
forces 26
  acting on body 26–7
forensic investigations 22–3
Frisch, Karl von 77–8
fruit fly (*Drosophila melanogaster*) 106–7

genetic manipulation 104
girder, compression and tension stresses 33
goldfish 61–2
gout 40
grasshopper leg 59, *59*
greyhound, movement of limbs 58, *58*

habituation 91
herring gull
  chick's begging response 70, 76–7
  egg retrieval 84, *84*
  model bills 76–7, *76, 77*
honey-bees
  communication 78, *78, 100–1*
  round dance *101*
  waggle dance *101*
hormones 97–8
horse, movement of limbs 58, *58*
hyaluronic acid 40
hydroid colonies 3
hydrotropism 3

imprinting 74, 92–3, *92*
inorganic phosphate 3
insectivorous plants 4
insects
  flight 66
  pleura 67
  tergum 67
  wing, attachment to thorax 67; bee *67*; crane-fly *67*; integument 67; movements *67, 68*; structure 67; vibration in forward flight *66*
insight learning 93
internal environment 97
invertebrates
  nervous system 85
  responses to light 87
ion movements 3
irritability (sensitivity) 70

jellyfish 3
jet lag 99
joints 39–41
  articular surfaces 40
  ball-and-socket *41*
  cartilaginous 39, *40*
  gliding *40*
  hinge *40*
  pivot *41*
  saddle *41*
  sutural 39, *39*
  synovial 40, *41–2*
  synovial membrane 40

kangaroo, counterbalancing *32*
kineses 85–6
klinokinesis 86
klinotaxis 87
kymograph 43, *43*

lactic acid 44
learning
  insight 93
  latent 91, 92
  trial-and-error 91
learning set 94
leg
  human, bones *36*
  relation to body weight 29–30, *29, 30*
  relation to stability 28–9, *28*
levers 45–8
  first class *45, 47*
  human body 47–8
  law 47
  second class *46, 47*
  third class *46, 48*
limb
  artificial 38–9, *39*
  bat *37*
  horse *37*
  human arm and leg *37*
  mammalian *38*
  mole *37*
  pentadactyl 36, *36, 37, 38*
  rabbit *37*
  vertebrate 36–8, *36*
  whale *37*
ligament 10
lignin 9
locomotion 1, 51–69
  human (walking/running) 55–7, *56*
  quadrupeds 57–9, *57, 58*
locust, courtship song 102
Lorenz, Konrad 73–4, *74,* 81

lovebirds *(Agapornis)*, behaviour patterns, 105, *106*

maturation 97
maze 91, 94
  learning by rats *91*
  pencil 95
memory 95
migration of birds 100
mosquito *(Anopheles gambiae)*, flight activity rhythms 99, *99*
mother–child relationship 80–1
mother-surrogates, cloth and wire *81*
motivation 97
motor end-plate 18, *19*
moulting (ecdysis) 24
movement 1–2
  drag effect on 4
  plant 1, *2–3,* 3–4
  principles 4–5
  thermonastic 3–4
mucin 40
multiple-flash photograph *75*
muscle 13–15, 41–3
  antagonistic 6; arthropod limb *59*
  cardiac *14,* 15, 42; nerve supply 42
  contraction 6, 18; myofibril changes *18*; nervous stimulation 18–19
  epimysium 15
  fatigue 44
  fingers, with/without tendons *45*
  perimysium 15
  sarcolemma (endomysium) 15, *15*
  sarcoplasm 15
  smooth 13, *14*; nerve supply 42
  striated (skeletal), 13, *14,* 16; coordination and control 42–4; motor unit 42; sarcoplasmic reticulum 17; T-system 17; ultrastructure 16–17, *17*
muscle twitch *43,* 44
myofibril 16, *17, 18*
myosin 17, *17*
myotome 61

nerve fibre, impulse transmission 43
nerve–muscle preparation 43, *43*
nervous system 84–5
  invertebrate *84*
neuromuscular junction 18, *19*

oddity problem 93–4, *93*
open field box 82–3
operants 79
orthokinesis 86
osteoblast 11, 13
osteoclast 13
osteocyte 11, *11*

Paramecium 53
  avoiding reaction 85–6, *86*
parasympathetic nervous system 42
patellar tendon 41
pecking movements 83
pheromones 101–2
phloem 7
photocell 75
photoperiodism 99–100
phototaxis 87
pineal gland 100
plants, environmental forces on 48, *48*
plasmosol 51
procuticle 24
prolactin 98
proprioceptors 41
pseudopodia 51

quadriceps femoris 41, 42
quadruped 57
  digitigrade 59
  plantigrade 59
  skeletal system, cantilevers in *33*
  unguligrade 59

rabbit
  forelimb and girdle *68*
  hindlimb action *57*
  stamping 100
Red Admiral butterfly 66
reflexes 90
  conditioned *see* classical conditioning
reinforcement 80
releasers (sign stimuli) 83
reproductive isolation 103
ring (Barbary) dove

(*Streptopelia risoria*),
reproductive behaviour
98
robin 83–4, 102
rodents, exploration of strange
environment 82–3

sarcoplasmic reticulum 17
sclerenchyma 21, *21, 22*
seaweed (*Ulva lactuca*) *23*
sensitivity (irritability) 70
sessile organism 1
sexton beetles (*Necrophori*)
72–3, *72*
sheep, gestation period 100
Siamese fighting fish (*Betta
splendens*) 107, *107*
sign stimuli *see* releasers
skeleton 4–7, 34–5
coral *5*
human *34;* hip region *34;*

prehistoric *5*
human body without *5*
hydrostatic 6–7
leaf *5*
materials *8*
Skinner box *79*
skull 13
sleep 99
snail, tentacle withdrawal 94
social behaviour 104
societies 104
song 102
stability 27–8, *28*
stickleback, three-spined
102–3, *103*
support system in mammal 35
swim-bladder 61–2
sympathetic nervous system 42
synovial fluid 40

T-system 17
tape-recording 76

taxes 87
tendo calcaneus *see* Achilles
tendon
tendon 10, *10*, 44
territory, male bird's 83–4
testicular development, birds
99, 100
testosterone 97–8
tetanus 44
tetrapod 57; *see also* quadruped
thigmonasty 3
thigmotropism 3
Tinbergen, Niko 76–7, 103
trajectories 31
tree, forces on 48, *48*
trial-and-error learning 91
tropism 3
tropomyosin 17
troponin 17
tropotaxis 87
turgor 7, 19, 20

typing test 94-5

vertebra *35*

walking
arthropods 59
horse *58*
human 55–7, *56*
quadrupeds 57
water, movement in 69
wilting 19
woodlouse (*Porcellio scaber*)
86–7, *87, 89*, 108
xylem 7, 9

yawing 61
young plants, turgor pressure
and support 19

zebra finches, courtship 108–9,
109